Building a
LOW IMPACT ROUNDHOUSE

drawing by Taus

Tony Wrench

Published by:
Permanent Publications
Hyden House Limited
The Sustainability Centre
East Meon, Hampshire
GU32 1HR
Tel: 01730 823 311
or 0845 458 4150 (local rate UK only)
Fax: 01730 823 322
Overseas: (int. code +44 1730)
Email: info@permaculture.co.uk
Web: www.permaculture.co.uk

THE QUEEN'S AWARDS
FOR ENTERPRISE:
SUSTAINABLE DEVELOPMENT
2008

First published in 2001, reprinted 2002, second edition 2008.

Designed and typeset by John Adams.

Printed by CPI Antony Rowe, Chippenham, Wiltshire.

FSC certified Evolution Satin 75% recycled paper used for textpages.

The Forest Stewardship Council (FSC) is a non-profit international organisation established to promote the responsible management of the world's forests. Products carrying the FSC label are independently certified to assure consumers that they come from forests that are managed to meet the social, economic and ecological needs of present and future generations.

FSC
Mixed Sources
Product group from well-managed
forests, controlled sources and
recycled wood or fiber
Cert no. SGS-COC-2953
www.fsc.org
© 1996 Forest Stewardship Council

British Library Cataloguing-in-Publication Data.
A catalogue record for this book is available from the British Library.

ISBN: 978-1-85623-042-1

The Author

Tony Wrench has spent many years designing and implementing, building and renewable energy projects. He lives with his mate Jane Faith in the community at Brithdir Mawr, in Pembrokeshire, West Wales. Their principles are sustainability, simplicity and spirit. Tony lives luxuriously, well below the poverty line, working on things permacultural and wooden. He makes his living from wood turning, singing and playing musical instruments (some homemade) with the local circle dance and Ceilidh band Rasalila.

It is as natural for us to build an appropriate shelter as it is for badgers.

Badger set under an old field bank tree at Brithdir Mawr.

Contents

Introduction ....page 1
Background 2
Design 4
Materials 10
  Embodied Energy 12
The Skeleton 15
  Reciprocal Frame Roof ..20
The Full Skeleton 27
  Inner Uprights 16
  Crosspieces 30
  Bracing 31
The Roof 33
  Secondary Rafters 35
  Willow Laterals 36
  Materials 37
  Roof Raising 38
  Eaves/Roof Edge 41
  Turfin' 44
  The Skylight 48
Walls 51
  Vents 55
  Cat Flap 56
  Straw Bales 57
  Platforms & Partitions 58
  Windows & Doors 59
  Basic Window Design 59
  Rubber Sill Sheets 62
  Lintels 62
  Back Door 63
  South Windows 64
  Front Door 66
  Stove Fitting 68

Water 70
  Grey Water 74
  Plumbing 75
  Solar Panel 76
Outside Touches 79
  Earth Sheltering 79
  Rubber Shingles 81
  Planting....................82
Electrics 85
Wood 89
Compost Toilet 93
  Construction 94
  Use 96
Inside The Roundhouse...98
Endpiece 103
Ten Years Later 105
  Strawbale Den 105
Roundhouse Feedback 113
Physical/design/lifestyle ...113
  Visual Aspects 113
  Cobwood ....,......115
  Roof 117
  Floor 119
  Compost Toilet 119
  Wood Store 119
  Renewable Elecricity 121
  Reed Beds ....,......122
  Kitchen 122
  Windows 123
  Design Lifestyle 123
Planning 125
Appendix A1

semper ut funquet

Introduction

Are you superstitious? I may be taking a chance writing this. By the time you read it our Roundhouse might have been demolished. It might have been listed as a historic building. It might be frozen in aspic as a Sustainable Lifestyle Heritage Exhibit with holograms doing repetitious sustainable activities in endless charade. It might have collapsed! It might have burnt to a cinder. Am I tempting fate by talking, nay, boasting, about it to you? Time will tell. *Permaculture Magazine* published an article about this place in issue No 20, with drawings by Olwyn, a friend who lives locally. Unwisely, I put at the end of the article: 'for further details write to Tony Wrench...' I expected one or two letters asking "Did you take the bark off the logs or not?" or "Did you use bolts on the sleeping platform frame?" or something. Instead I keep getting letters just saying 'please send me more details'. OK, I give in. This book is the more details. Not everything. No fashion tips, recipes or laundry hints for hovel builders. I'll try and put enough details in without boring you silly. Caution: I am not even a cowboy builder. I'm not an architect or qualified to build anything. This place has more in common with a shack in a shanty town in Buenos Aires than it does with a new Wimpey house in England. Don't expect any expertise from this book. If you build something like this roundhouse and it collapses, don't blame me. This is about doing things that are not regulated and predictable, I'm afraid. Naturally, I'd be delighted if you want to build a roundhouse and use some of the ideas here, but if you're worried about safety, consult an architect first. If you are a natural worrier, better treat this as a bit of semi fiction, from another place and another time.

Background

Jane and I have lived in West Wales since Dec 1989, trying out the ideas of permaculture on a smallholding of 1.5 acres, growing our own veg., keeping chickens and ducks, living simply. We experimented with ways of reducing our bills by giving up a flushing loo in favour of a compost toilet, and by building solar water heating panels and a rainwater shower heated by the sun. We converted the house lighting to 12 volts, installed photovoltaic cells and a small wind generator, and I built and played around with various designs of den and studio buildings made from recycled materials (especially windscreens and double glazing units) and slab wood offcuts from the local sawmill. We got rid of our car and bought an electric milk float. A normal country life, in fact. Each week we would hold a practice with Rasalila, the acoustic circle dance band we are still part of, in the Cone, a wooden tipi studio that doubled as our bedroom. Emma, who plays fiddle, bagpipes, whistles and clarinet, used to fill us in on progress at Brithdir Mawr. A tumbledown farmhouse and outbuildings with acres of fields and forest about 18 miles from us, that she and her husband, Julian, had bought with the intention of forming a sustainable community. We would exchange experiences about our respective gardens, and Emma would often remark how easy it would be if we built a little den down near the woods at Brithdir. I could still do woodturning there, and she would no longer have the long drive to band practices. As we saw the community beginning to take shape over the next two years we realised that these suggestions had a lot of sense in them. It's OK living in an agricultural workers' bungalow in Wales in the summer, but each winter you become aware of just how badly they are designed and how poorly insulated they are. I've lived in four since 1977, and not a year has passed without me thinking about designing a house of my own that wasn't always cold and draughty; a house that faced and welcomed in the sun; a house that made sense and was fun to live in; something that didn't cost the earth to build and didn't need a mortgage, so left me free to live simply. Jane and I had 1.5 acres of tightly packed Douglas Fir alongside our track. One day we started thinning the wood. We were going to build a den at Brithdir.

So many people would love to build their own place, and so few do, especially in Britain. So many need to live and work and be in nature, yet so few are allowed the space to do so. I can understand the planners' fears that if everyone were allowed free range to build what they liked where they liked this small country would be overrun, but yet there is something in me that revolts against a system that assumes that I and nature don't mix. At the heart of our planning laws is the unspoken assumption that people and the countryside are bad for each other. This is clearly nonsense - we have only evolved as we have over hundreds of thousands of years, by being tuned in, precisely and acutely, to the seasons and rhythms of nature. It is as natural for us to build an appropriate shelter as it is for badgers. There is no evidence that it works in the long term to keep humans confined to towns and cities, except for a few mechanised industrial farmers who are supposed to keep us all supplied with cheap food. It's not working for wildlife, for the farmers themselves, nor for us, fighting as we have to for overpriced and badly designed flats and houses in fume filled towns and cities. For us to live sustainably on this endangered jewel of a planet it will take radical re-appraisal of how we relate to and use the land, and I want to be part of the solution, not the problem.

We had to think all these things through before deciding to build a den that would certainly have been denied permission, had we asked for it. At the moment (and things could change very quickly) I consider it folly to assume that the powers that be are capable of taking rational, sustainable, holistic decisions. They are too stretched; their offices are too noisy; their journey to work is too stressful; they are too cut off from nature. I worked in that world for seven years. A council Chief Executive once said to me "I'd love to be green, but I can't afford to. I have to fiddle while Rome burns."

I don't want to make anyone wrong. We're all doing our best. I've simply decided that no one has the motivation to sort my world out more than I have. I have taken reponsibility for my life, so I decided to build my dream house here, before I get too old to be able to lift a rafter. Anyway, it's in my genes. My grandfather built his own wooden house. Now I've done it too, it seems the most natural thing in the world.

Design

On with the nitty gritty. I have realised, through making plenty of mistakes, that it pays to think hard before building anything, and to write down my thoughts as I go along. I have found the principles of permaculture (Pc) an essential tool in this design process, and cannot recommend highly enough that you go on at least one Pc design course and read at least a couple of Pc books before you consider building your own home (or even a home for your hens). I started this project by re-reading my favourite book on self-build, Ken Kern's *Owner Built Home* and his sequel *The Owner Built Homestead*. Then wrote a simple brief to myself: "Goal: An autonomous house of wood; very warm, very dry, cheap to run. Made from pine logs from Erw Deg (where we lived). Turf/bracken roof...It is built on a slope near woods."

Over the next year, we agreed the best site for the house with Emma at Brithdir. It had to be preferably on a south facing slope, invisible from the mountain, Carningli, or surrounding hills and near the woods. The place we chose was on a bank covered with bracken in one of the smaller fields, just below a pair of hawthorn trees, near the end of a green lane and about 400 metres from the farmyard and main house of the community.

Brithdir Mawr and Carningli mountain.

4

The whole farm is 165 acres, rolling fields, big hedgerows with oak and ash standards, and includes about 60 acres of woods. Big and green and fairly wild. So I needed to take some time thinking about an appropriate building/den/ hogan/cabin for this kind of landscape. Something that would not impose anything on the land or be inorganic and heavy. Something that could even rot away when we are dead. Something incorporating low

Iron age dwellings at Castell Henllys.

embodied energy. Lowest building materials for embodied energy are cob, straw and unsawn wood from on site. Also, cement manufacture now accounts for 10% of global CO_2 emissions.* OK. No cement. This was going to be an attempt to build a home whose materials were very natural and very local. A sustainable home.

We had two young people, Dima and Carol, staying at our place as 'Wwoofers', working for their keep for a couple of months. For one winter (96/7) we spent at least half of the daylight hours available thinning our bit of Douglas Fir wood across the track. Each day we would note the trees that were too crowded, too small or that had been blown at an angle by the wind. With an axe and bowsaw Dima and I cut them down and Jane and Carol did the snedding - cutting off all the little branches with hand axes. In the evenings I read *Shelter*, the American book on ethnic hand built housing the world over, and borrowed a book on log cabins. I revisited Castell Henllys, two miles from here, where there are precise reconstructions of a settlement of round wood, mud and thatch houses from two thousand years ago, rebuilt in their original post holes; all made of natural materials.

The design for this roundhouse started out as an eight sided hogan with split level roof whose centre would be supported by enormous forked beams. The roof would be Navajo style 'whirling logs' - a circular

* *source Tomorrow's World.*

pile of 100 trees. The walls would each be 12 ft (3.5 ish metres) long, joined at the corners by overlapped joints. The log cabin book showed how to cut each log along its length with an indentation in its underside so that each would fit tidily and warmly on the one beneath it. This seems the best alternative to adzing each one roughly square, or having the whole lot milled by a sawmill, a prohibitively expensive operation. I tried it using an electric chainsaw. The little machine just couldn't handle the work and I couldn't handle the noise and strain of shaping every single wall member along the grain like this. OK, back to the drawing board. Where was that tiny reference to cordwood walls - using different thicknesses of logs, all cut to the same length and stacked up as in a firewood stack (or 'cord') to make a wall? I found a couple of small references. It was a technique used by European (usually Swedish) settlers in the mid-west US and Canada in the 19th century, to make 'poor people's' houses. Warm, easy to build for a non professional, and using all sorts of wood. The book also said that softwood, such as pine, loses heat two and a half times more readily through the end grain as laterally. To have the equivalent of a 6"/150mm thick wall - a fairly standard Swedish style log cabin thickness - I would need 2.5 times that as cordwood, i.e. 15" logs. I actually settled for 16", to be on the warm side, and because it translates easily into metric - 40cm. OK it's a weird reason, but it's the truth.

If you ever get into designing things, you will find that you evolve ways of working that make sense to you. You have to, because it actually involves a lot of concentration. I have two basic methods. The first is to carry out a conversation with myself in writing and drawings as the design takes shape. I have 25 - 30 pages of these pencilled conversations showing the gradual evolution of this design over about nine months. It is useful at the end of a design session on, say, the roof, the floor, the windows, the drainage system - whatever - to state the conclusions drawn and the things to do next. This is because all parts of a holistically designed building impinge on each other, and often you are left with questions rather than answers. For example, I left any detailed consideration about what to make the back walls and floor with until the JCB had dug the hole into the bank. This was because if we had hit rock a foot down the whole design would have needed to change. It was worth writing down the questions I was left with after each design session. Doubts too. I have long harboured dreams of heating a house by the slow convection of air from pipes through and around a vast underground water tank that has been accumulating heat from solar panels all summer. Fair enough, lets go for it. As I worked on thinking it through it became obvious that this was actually going to be more complicated than I could handle. Maybe you've got plenty of money to pay consultants for this kind of thing. Fine, but we had no money at all. This whole house was built on trickles of cash in, using natural and recycled stuff, and good luck. So if there was a doubt that (*a*) the water and plumbing system would be just too complicated and (*b*) a huge water tank in the middle of the floor might affect the footing for the main house supports (just a feeling rather than anything definite), I chose simplicity and went back to the drawing board again.

I am sorry if this seems too long winded for you, but I cannot stress strongly enough that if your house is going to work for you perfectly in the situation you have, then you will need to take just as long designing it as building it. So much waste in our society comes from people wanting a quick off-the-peg solution, and that usually means more transport, less sympathy with the environment, and just silly design. Look at all those new housing estates with the houses dutifully facing the road, whether the road is to the north or the south. Simply by locating a house to face the south you are reducing heating bills by at least 13%. And all those south facing roofs with not a solar panel or PV tile to be seen. It drives me mad. Planning? Don't make me laugh.

The other type of design work is something that maybe you already do very well, but I've never seen it talked about much. Virtual Reality is now a concept that many people are familiar with, though, so it is certainly happening in cyberspace. I hope the Kogi native people of Colombia will not be offended if I use their phrase 'working in Aluna' to describe it, for that is what I call it. The mind is a wonderful thing. Recently psychologists advising an American football team carried out a series of experiments on the team members. Success in American football apparently involves a team in having a series of multiple-move strategies that the captain can summon up, like a program, in the course of play by yelling out a number. The team then make the appropriate moves to carry out this strategy. The psychologists worked with the team to go through several virtual games, hunched up together in a room without touching a real ball, concentrating in their minds, imagining each

7

move and the counter moves of the opposition. It was assumed that this kind of virtual practice session would not be as good as the real thing, but might just be of some use. In fact they found it to be of more use than a real practice, and got better at it, too. The team performance dramatically improved. When I read this I remembered the film From the Heart of the World that the BBC made at the invitation of the Kogis, a native nation in Colombia that has retained its

Rob Roy's cordwood and masonry house.

culture since before the conquistadores. As I remember it, the Mamas, the leaders, do not take any decisions on the physical plane until the consequences and details have been worked on in Aluna; in darkness, in their minds. Using this idea, I developed for myself a way of imagining in detail everything about a particular aspect of the house that I was working on. If you do this you will find that you can at first only imagine one or two steps or objects, but as time goes on you can develop Aluna into a very creative space in which questions can be posed, practical solutions put forward in succession, and each possibility looked at in 3D and full colour. You can turn something around, imagine it in pouring rain, imagine five nine year old boys climbing all over it and so on. As this house became a reality, I found that at least half of the time I was working in Aluna, usually on a practical detail that I had never actually encountered in real life, or at least never on the same scale. For me the best time to do this is the middle of the night, unfortunately, when all is totally still, so Jane got quite used to me getting up at two or three a.m. to draw some detail of the plumbing system, a window frame or the roof rafters. The Kogi, I think, believe that on the Aluna level all minds are linked. I have since found out that several ideas that came to me in Aluna have been the subject of quite detailed trial and error work by pioneers in alternative building methods such as Rob Roy in the US, who contacted me when my house became known about and told me that it was a cordwood masonry house.

Anyway, there you go. It is a way of designing that soaks in all available data and then allows another part of the mind to mold it. It works for me. It is the main reason why I couldn't apply for planning permission in advance, as a matter of fact. How can you put in detailed drawings if you don't know what you are going to do until the night before you do it? From now on there will be a simple description of construction details. Please don't ask me to explain everything, though, because I've forgotten most of it. We just woke up the next morning and got on with it. I bet that's how our ancestors worked for the last few hundred thousand or so years that we have been building simple shelters.

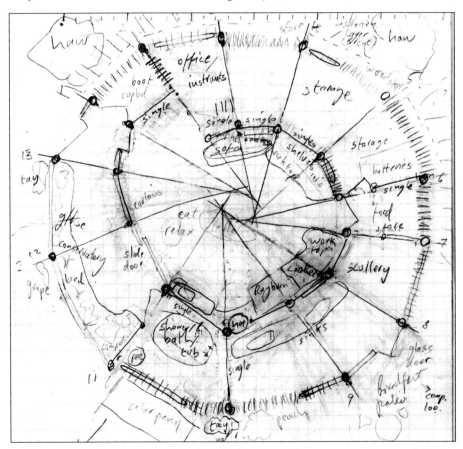

(diagram 1) This is about as detailed as any design got: post and beam structure, round, with reciprocal frame roof. Outer circle of uprights to go up first, then henge crosspieces, then roof. Cordwood in walls.

9

Materials

We took a winter to get the wood cut. Dima, Carol, Jane and I would plunge into the damp deep sloping pine woods in the mornings with a sharp bowsaw, two little axes for snigging (I think that's an old word for cutting off minor branches - the other good word like that is snedding, which is dragging the big logs out by horse power), a really sharp felling axe, some rope, and tough gloves and boots. The inventory I was working to was:

roof supports (rafters) 6m+ - 13 off plus spares

pole uprights (9" diam pref) 3m - 13 off

inner uprights (8" diam) maybe 5m - 13 off

walls: a rough guess of 210 x 16" (40-cm) logs per full wall, = approx 24 lengths of about 4 m each! Say 9 walls max, allowing for windows, so: lengths of all thicknesses 4 m - 200 off.

So we had plenty to be getting on with. We got really good at tree felling and thinning by hand, and measuring by foot. I used a tipi pole marked out in lengths of 4m and 6m with yellow insulation tape, but the working situation was temperate forest with brambles, small streams and occasionally almost impenetrable undergrowth at a slope of about 1 in 3. I use metres for stuff like this because 1 metre is one of my paces with big Swedish army boots on (vintage 1941! - aren't street markets amazing sometimes?).

Thinning trees for building material in our pine woods.

Logs being transported by milk float.

I like to have all important building materials ready before starting work, so made a call to Kevin Thomas, a friendly glazier whose workshop was only 10 minutes away by milk float. Glaziers get enormous amounts of double glazing units that are perfectly good, but have been taken out in some renovation job, or have been ordered in the wrong size for a particular frame. If you can be flexible in your size requirements for windows, as of course you can with cob, wood, cordwood and strawbale buildings, you save a fortune on windows. One day Kevin called me back. "Got a load of patio doors and windows. I'm clearing out the workshop. I've got to skip a whole load. Do you want to come up in the next half hour (sic) to have a look?" SCRAMBLE. I bought three double patio doors, about three ordinary wood framed small windows, and about 15 big double glazed units (several too big for one person to lift) for £125. Just fitted them all on the open back of the milk float. (Don't ask me all about that - suffice to say that if you want to wean yourself off a car, get a second hand electric milk float. The old batteries will still last you about three years. You've got a range of about 25 miles for an average 15 electricity units, charged overnight, and a float will carry 1.3 metric tonnes. That's a real load of wood, logs, manure, or windows. Mind you, keeping one going requires a steep learning curve involving DC electrics and battery charging and maintenance. If you're tempted though, check out your nearest dairy that uses floats, and go for it. Make all your neighbours laugh.)

The only other materials I bought in advance were 2 rolls of galvanised steel fixing strap from a builders' merchants, and a good supply of nails of all sizes, including 30-40 6"/150mm nails.

Embodied Energy

This is the energy tied up in a structure by virtue of the transport, extraction, processing and other costs that go into all the building materials. Cement is very high in embodied energy. So is aluminium - all that mining, all that electricity to make it; transport from, usually, Canada. If all countries in the world used as much aluminium by 2050 as Britain uses today, the global capacity to sustain such production would be exceeded by eight and a half times. New glass and bricks (all that baking) are also high in embodied energy. Generally, the true cost of road transport, if built into the embodied energy costs of building materials, can make a huge impact on the environment. If you use wood in London that was felled in Scotland, for example, it will have a higher embodied energy than the same kind of wood imported from Latvia, because the latter comes by boat to London docks, whereas the Scottish wood has come maybe 600 miles by road. This is one of the critical, yet almost invisible, factors that form part of the big issue of living sustainably in everything we do. Seeing just how much wasted energy goes into the building of an ordinary modern house was one of the main reasons that drove me to designing and building an alternative model from scratch. I learned most of it by joining the Alternative Technology Association run from CAT (Centre for Alternative Technology), the Ecological Design Association and reading books like *Our Ecological Footprint* and the FOE publication *Tomorrow's World* (*see Appendix*).

The house with rock bottom embodied energy, and therefore the kindest to the earth to build, would be made of natural materials found and processed by hand on site. Our ancestors built houses from cob (clay with some sand and straw), local stone, wood and thatch. Nowadays I would contend that by adding straw bales and a modern technological miracle - large seamless rubber pond liner sheets - to this list (admittedly rubber is imported, so higher in embodied energy, but grown in sustainable conditions giving livelihood to rainforest dwellers), we can still produce warm, dry and organic houses at a fraction of the cost to ourselves and to the environment.

We brought the wood the 19 miles from our old place to this in two big lorry loads, the lorry being specially equipped with a

This is the amount of wood needed for the structural framework.

long lifting arm and chains. We then carried it the 400 metres down the track by tractor and trailer; maybe twenty loads, spread over the next three months or so. We used no power tools on site, so I cut up the wall log-ends up by the farm house with an electric chainsaw. (I'll follow American custom and call them log-ends from now on). So even a low impact building like this has quite a bit of embodied energy. Electricity was consumed for the cutting. Some was renewable energy, mind you, because our community bought an inverter that enables most electric appliances such as hand tools, electric chainsaw, computers etc. to be run from the battery bank fed from stream turbine, solar PV panels and wind turbine. Diesel was used for maybe 80 miles of lorry transport plus maybe 5 miles of tractor use. I have some regrets that even the small embodied energy costs in this house could not be avoided, but it is pointless being against all technology, isn't it? We didn't have to build an earth sheltered house into a bank, but I'm glad we did (there's a great gale blowing around outside as I write this and we hardly notice it in here) and I'm grateful a JCB was on hand to save us weeks of back-breaking digging into clay.

The JCB revealed the perfect material for filling between the log-ends - cob, as we had hoped. If we had hit rock a foot down I would have had

Inside the bender (inset) the outside view.

to redesign the whole thing, so flexibility was crucial. We made three piles - subsoil to either side of the hole and one big pile of topsoil for spreading back around the house and for the roof.

By September 1997 we had the main logs by a big clayey hole in a bank. Jane and I set up a bender at the top of the bank, made of bent hazel poles covered by canvas, and borrowed a little wood burning stove. This was our home for four months. Don't attempt this with someone you don't know very well. At times, after weeks of rain, mud and sleet, it felt like Scott of the Antarctic's last camp. Slugs eating our books; wheelbarrowing batteries through ankle deep mud; wet clothes, aching bodies. I won't go on any more about it, but please don't think that building your own place is some kind of easy option! Right. We're ready to start.

The Skeleton

A traditional and safe way of building a simple house is to make a 'post and beam' skeleton and fill in the spaces. The skeleton of this roundhouse is a wood henge of uprights with cross pieces forming a complete circle, and a reciprocal frame roof, all of round Douglas fir poles. The reciprocal frame means that the rafters rest on each other in a circle at the centre, where there is a hole. More strength, for a turf roof, is supplied by an inner circle of supports, but the outer circle comes first, and the size of that was actually dictated by the flat space we had available when Berwyn left on his JCB.

First we laid out the rafters on the ground to get an idea of the thing. It's no use building something to some theoretical size if in practice the ends of your rafters are a bit weak and dodgy looking. It's always worth having something small and strong rather than maximum width and sagging in the middle. So Willow (a permaculture friend and tree planter extraordinaire) and I laid out the strongest of the poles around the circle. "How many poles you having?" asks Willow. "Twelve," say I, because I've already made one with twelve poles in the field across the way, that we still use for a marquee frame. "No," says Willow, "It's got to be thirteen. I'm not going to help you build a house with twelve sides. It's got to have thirteen." It's no point asking him why. Willow can talk all day about the moon phases and the Mayan calendar and stone circles and just about everything. In a quick mental resumé of the last nine months of design work I think, "How attached am I to it having twelve sides?" The answer comes back - not at all. The more sides, in fact, within reason, the better, as each cross piece holding the sub-rafters has less weight to hold. "OK, thirteen," I say, and from that moment on all thought of attempting to calculate measurements in advance goes out the window. Try dividing anything by, or multiplying it by, thirteen. From then on it was rule of thumb. Just as well really, because I didn't have a tape measure long enough to reach across the whole house circle, and still don't. The circle is not exact - it fits the space and suits the strength of the rafters, but I still can't tell you precisely how wide it is.

We laid out the thirteen rafters and stuck a stick in the ground where each support would be. Then we started digging holes and painting creosote on the bottom metre of the main supports. I know that poles rot if you put them in the ground, especially in the area just around ground level, where bacteria and fungi have moist air and damp wood to work on. Oak and chestnut resist rot for 40-50 years; sycamore, hazel and ash will last about 18 months, and most woods are somewhere in between. On previous projects I have tried charring the bottoms as our ancestors did, creosoting, standing the pole in a plastic bag with creosote in and putting concrete around the bag, and doing nothing. After about ten years, none have rotted away yet, but I guess the 'doing nothing' will go first. If we were using concrete here, we could stand the poles on concrete plinths, steadied by stainless steel pins, but then the uprights would need much more bracing than they've got, to stop them twisting or moving sideways. Call me old fashioned or unrealistic but I don't think you can beat a nice hole with the pole stuck good and fast in it. This whole building will rot anyway, one day. I don't want the poles to rot faster than anything else, though, so each time I clean out the chimney I put the natural creosote and soot into the clay around the base of a different pole. The holes are a cubit deep. Here's Jane finishing one off. When she can't scrape any more from the bottom of the hole, that's a cubit.

Jane digs a post hole; she can just reach a cubit down.

When all the poles were in, we cut them so that the tops were all exactly level. This is important so that the whole place doesn't look completely skew-wiff. The load from the roof is distributed evenly, and I could use the henge cross pieces as a level when it came to fitting windows, etc. How to do it is to make a bunyip. Not a word you see much, is it? A bunyip is a length of hose of approx 10m length with about a metre of clear plastic

The posts go into position.

at each end. If you fill this with water, making sure that all air bubbles have left, you will find that the level of the water visible in the clear plastic ends is the same, no matter how far apart you hold the ends. Don't attempt this without a friend helping you. You nearly go mad as it is, because it's all too easy to move one end up or down too suddenly, with the result that a jet of water comes from the lower end, and you have to replace the water and start again. Still, when you get the knack of it, which can involve marked sticks attached to each end, pieces of chalk, communications skills and a bit of cursing, you end up with a series of marks on the uprights that you can guarantee, by scrupulous cross checking, are all level with each other. I started with the pole in the south west that I knew I wanted for a door post, and made a mark 2m from the ground level. (The other time you may find a bunyip useful will be if you want to dig a swale or ditch for draining/water retention along a contour. In this case a marked stick against the clear plastic is essential.).

At this point we have 13 marked uprights in holes. I actually cut the tops off as they were standing, as I could not manipulate them single handed (each pole was about three metres long and at least 25cm in diameter, so weighed more than me). If there is a team of you, you can take each pole out, cut it to size then replace it, but I find earth falls in the hole and I can't be sure of the level again, so I cut them where they are.

Each pole stands on a rock at the bottom of the hole, is jammed tight with stones, then tamped upright using a plumb line. Don't try and put up your main supports without testing that they are plumb straight. No point tempting gravity, is there? There is a case for drilling the hole in the top of each upright before you do the final tamping, because the pole moves a bit as you lean against it with a step ladder. Drilling these holes, 3 - 4" (75-100 mm) into the end grain 2m up a stepladder, with a brace and bit, was the single most exhausting operation of this entire project, by the way, so when you've done that, relax. The rest is downhill.

To make the henge, I measured the exact distance between the centres of each upright at ground level, and cut the cross piece to this length plus one pole diameter. This is important so that your henge pieces overlap and are pinned together on the top of each upright (see diagrams 2 and 3). This is a way of doing it that I have evolved over ten years or so. If you can do it better, by all means do so. It is important to use the henge to hold itself as a strong circle, to allow for most of the roof weight to come down each support evenly. The pin going from the upright through the cross pieces into the rafter ensures structural strength, and acts as a useful fulcrum when adjusting the rafters. At first it is a fiddle cutting each cross piece to get the angles reasonably tight, but actually it doesn't take that long, and looks great. As the electric chainsaw would save a lot of time with these overlap joints, I cut the cross pieces in kit form up by the community workshop, using measurements of each gap marked out on a piece of paper. I have fond memories of this scruffy, damp piece of paper being passed from jacket to trouser pocket and back with the circle of posts marked out and 223, 219, 226 etc. in all the gaps. (I find cms are quite good for measuring this kind of size, then feet for a bit bigger, then metres for long lengths. What do you estimate and measure lengths in? Be honest. It's a mess, isn't it?)

I had the henge finished in two weeks, working every day, usually with Emma as helper, to whom I am eternally grateful. At this point I could see what the size would be - the henge contains the space, and the imagination starts filling in plenty more ideas.

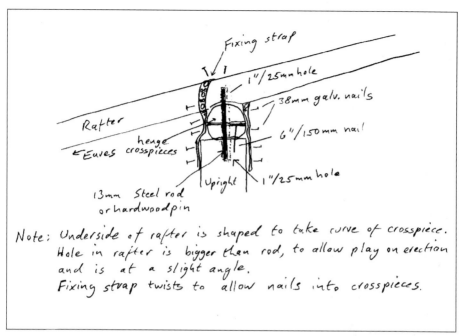

Fixing strap

1"/25mm hole

38mm galv. nails

Rafter

henge crosspieces

Eaves

6"/150mm nail

13mm Steel rod
or hardwood pin

Upright

1"/25mm hole

Note: Underside of rafter is shaped to take curve of crosspiece.
Hole in rafter is bigger than rod, to allow play on erection
and is at a slight angle.
Fixing strap twists to allow nails into crosspieces.

(diagram 2) Henge, crosspiece, rafter fixing detail.

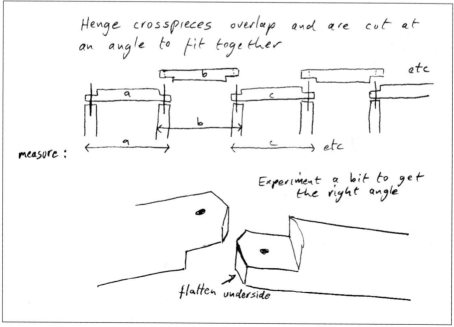

Henge crosspieces overlap and are cut at
an angle to fit together

etc

b

a

c

b

measure:

a

c

etc

Experiment a bit to get
the right angle

flatten underside

(diagram 3) Arrangement and joint detail for henge crosspieces.

Reciprocal Frame Roof

Ever since building my first reciprocal frame roof, on a simple building at our last place that we called the Dojo, I have been fascinated by the ease and strength of them. The idea was made popular by an architectural firm called Out of Nowhere, but I don't think they invented it. Our friend Jack Everett built his dojo near Stroud with one in the early 1980s. Anyway, it is a great way to use round timbers to hold a roof up without needing a central pillar; you get a hole for a skylight instead. The essence of a reciprocal frame is that the rafters are self-supporting at the centre; each resting on the one under it, in a circle. You do not attempt to hold the rafters at any particular pitch - you let them rest on each other and the pitch is what you get. The weight is transferred evenly down each rafter to the uprights, and vertically down to the ground. Putting up a reciprocal frame is great fun, but has an element of danger too, so it's best to ensure that there are no children running around when you do it, and that if you have helpers, they are not afraid of long poles rolling about above their heads. (Some people are - surprising, isn't it?)

Now we have a henge of posts and crosspieces, with a hardwood peg projecting about 20cm above the joint. The essential extra for erecting a reciprocal frame is the Charlie stick - a strong pole in a Y shape strong enough to hold all the rafters up until the last one fits in above the second to last and under the first one, when the Charlie stick can be removed. I have a rule of thumb with these roofs that seems to work: the length of the *Charlie stick must be: henge height + (no. of rafters x their average thickness at their ends) + 1 foot/30 cms, minimum.* For this house the Charlie needed to be at least 14 ft; that's about 4.5 metres long. Such poles are not easy

Dojo at Erw Deg.

The nearly completed henge, with a crosspiece awaiting fitting.

to come by, so when you find one, mark it, tell all your helpers about it, and don't cut it up. From the 250 or so trees we felled I found one, and only one, perfect Charlie stick. So now, a year later, its time had come. First I built a stepladder tripod like an adventure playground climbing frame about 4 metres high in the middle of the circle, using the longest step ladder we had. This actually felt much safer to climb on than the step ladder itself, as the base was well puddled clay, and the step ladder with its neat little metal feet would gradually tilt to one side until it fell over.

Next I set up the Charlie stick in its place. This needs some thought. The best way to get an even circle hole/skylight in the middle of your reciprocal frame is to have a standard alignment of each rafter pole. If you want a clockwise spiral/iris effect when you are standing under your roof, as I did, then you need to move clockwise round the circle in placing your poles, and each pole has to be displaced a standard amount to the right of the centre of the circle. If you are using an even number of henge uprights, this is a matter of standing at one upright, looking at the pole directly opposite you, then aiming at, say, the centre of the space to the right.

A point near the next upright will give you, ultimately, a large central hole - a point near the centre line will give you a smaller hole. In this case, as we had an uneven number of uprights, I aligned each post near the post on the right of the space opposite (*diagram 4*). This means that the Charlie stick, which has to hold the first pole, must itself be off-centre. I wanted a hole about a metre across, so figured that if I offset the Charlie by half a metre to the right, that would work, and so it did. The Charlie was put up standing vertically on the ground, gripped in the jaws of a workmate to stabilise it lower down, and lashed to the wooden tripod to stabilise it high up. It is good if you can also have an assistant to help you to hold the Charlie stick as the first rafters are being put into place, because if it collapses half way through it's like Pick-up-Sticks with hundred-weight heavy, 6 or 7 metre long sticks. I got all the best poles ready, fanning out from their uprights, and drilled a 28mm hole in each about 3"/75mm deep, at a slight angle, and allowing at least half a metre or more for the eaves. I also made a slight notch in what would be the underside of the rafters to allow a good sit on the henge (*diagram 5*).

It is at this point that good pole selection pays off; each pole must be thick enough at the centre to be capable of holding a considerable weight at its thinnest part, where it meets the others at the centre. If a pole is too long, you can always cut off the surplus at either end, so in this process it is always best to use good strong poles that are too long. The surplus length at the eaves is actually quite useful, as you can manipulate a pole much easier around the fulcrum of the henge if you have 2 metres or more of heavy end to play with. Several of my poles had this much spare, so I drilled the pin hole accordingly further from the end. The outstanding ends were again useful, when we clad the roof, to clamber up on to the roof and to lean secondary rafters against. The houses built by our ancestors from cob and thatch, such as the ones at Castell Henllys, often had huge spacious eaves coming down almost to the ground. These would provide dry cover for animals, firewood, hay etc. and would of course protect the outside wall from the elements. These days, long rafters would also provide the basis for greenhouse extensions in the sunny sectors. Between you and me, I have often found our generous eaves also of use for keeping dry while peeing in the rain.

Right, so our rafters are drilled, and ready. The Charlie stick is waiting.

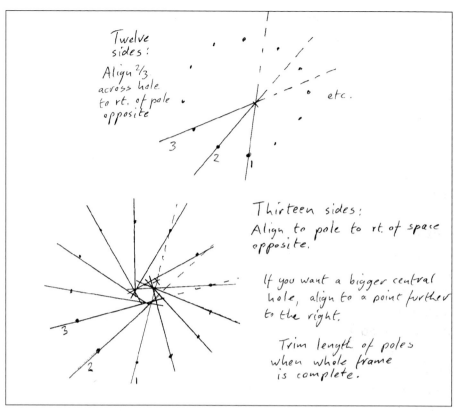

Twelve
sides:

Align ⅔
across hole
to rt. of pole
opposite

etc.

3

2

1

Thirteen sides:
Align to pole to rt. of space
opposite.

If you want a bigger central
hole, align to a point further
to the right.

Trim length of poles
when whole frame
is complete.

3

2

1

(diagram 4) Reciprical roof frame rafter layout and placing sequence.

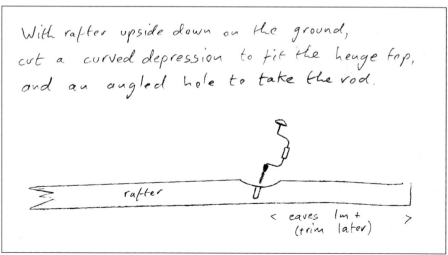

With rafter upside down on the ground,
cut a curved depression to fit the henge top,
and an angled hole to take the rod.

rafter

< eaves 1m +
(trim later) >

(diagram 5) Rafter detail for fixing at henge crosspiece.

I had a spare, smaller Charlie of about 2.5 metres long (let's call it Ajusta) for adjusting the rafters once they were up. So I lifted up the first rafter (in the South; no reason, but Willow would probably approve), heaved the thin end up onto the henge, climbed the tripod holding the end of the pole, and, on tiptoe, lifted the first pole up and into the Charlie Y. I was not actually being assisted that day, so that ten minutes or so was the nearest I have ever got to being a circus performer. I got a bit of birdsong for applause, and tied the pole loosely, to allow some movement, with baler twine. (Baler twine? We use it as currency here. I once assumed it's minted somewhere and planted by God in hedgerows and on old farm machinery the world over, but find that you can buy it in big rolls from agricultural Co-ops.) I then returned to the ground, lifted the heavy end until the hole was over the peg in the support, and let it down. With each junction at the henge, I use a piece of builders' galvanised steel fixing strap, which comes in 5 metre lengths. I cut off about a metre, and half fix it with a couple of nails on the upright, then one nail on the top of the rafter. That stops the rafter rolling anywhere, which is what they sometimes try to do. Stop to take a photograph. Note the pegs sticking up from the henge, the Charlie stick and the tripod.

Henge, tripod, Charlie stick and first rafter laid in position.

From here it is logical and straight forward, going round the circle adding each rafter, aligning each to the appropriate place across the circle, adjusting each one at the centre with Ajusta, tying with baler twine and half fixing with fixing strap.

If you are likely to be building one of these yourself, may I recommend that you do this process with a scale model of sticks first, say tenth or one fifth scale. There is no substitute for experience in

1996 field marquee structure.

making the small adjustments of each rafter as it goes on so that the central hole becomes regular, and it is much better to get the hang of it with, say, 1m sticks than with giant 7m ones. Having said that, I still believe this system is perfect for people with virtually no building experience, so long as you have played with long poles, levers etc. a bit. Now we come to the last two stages of the reciprocal frame skeleton, which are The Last Pole and Taking Charlie Out.

The last pole fits under the first and over the second to last. If Charlie is too short there will be no room for the last few poles. If he's too long the poles will keep trying to roll away down each other and Taking Charlie Out will be a bit dramatic. With Charlie at formula length I had just enough space to squeeze the last pole into its space and lower it down onto the peg with a bit of grunting and groaning. It is now just a matter of removing Charlie. Of course Charlie, by now, has come to realise that he, and he alone, has been holding up about half a ton of woodwork, and is indispensable to the entire project. He is well rooted in thick clay. Removing him feels hazardous, even if it is not. I will always remember when we were at this point putting up our field marquee structure in 1996. One of the observers, wringing her hands in dismay at the apparent certainty of imminent loss of life, wailed, "What does it say in the book?" Well, here's the book, four years late, and it says Don't Panic, take it slowly,

25

and dig down under the Charlie gently until, bit by bit, the structure takes the weight, the poles settle, and Charlie can be removed, preferably to carry out a ceremonial function somewhere. I adjust the circle with Ajusta, then finish pinning down all the fixing straps across the tops of the rafters and down to the top of the upright on the other side. I then fixed the poles at the centre with short lengths of fixing strap on the top sides of the poles, and by my favourite fixing system - a metre of telephone wire wrapped round several times and stapled two or three times. The basic skeleton is now complete.

Detail of telephone wire strapping around the skylight hole.

The Full Skeleton

Turf roofs are heavier than almost any other kind of roof. If your house is to be built in a region where winters bring heavy snowfalls, you need to estimate how much strength you need in your roof to hold up not just a thousand or more wet turves, but also a layer of snow. Heavy snow falls are very rare here in this part of West

The roof with a typical light snowload.

Wales, so this roof is not built for the worst case scenario. I also did not intend to use turves much thicker than about 3.5 inches, or 9cm, as it rains often enough in the summer to prevent the turf totally drying out. Still, a circular roof 12m/40ft across gives me approx 1400 turves of one square foot. Each turf weighs from 1kilo (2.2lbs) to a max of 10 kilos, averaging about 5 kilos, so for this roof we are talking about something like 7 tons of turf when wet, plus insulation of about a ton. Looking again at accounts of ethnic round houses, I noted that usually there were inner posts to bear the extra weight of a thick roof. The circular earth lodges of the early British, and North American Native peoples, the Mandan, the Pomo and the Miwok, all used an inner henge of strong poles to support heavy roofs. This house therefore also has an inner henge of strong uprights and crosspieces to support the load-bearing rafters near their mid point. If you are worried about load bearing capacity of roofs, there are experts who can calculate these things, so consult one. My basic rule of thumb with this house was: *(a)* to build with the strongest members I could actually lift, because in this site and weather it would be silly to have always to rely on one or more extra helpers all the time; *(b)* to use experience of building with this type of roof on seven or eight previous structures; and *(c)* to trust the designs of our ancestors who built and refined roundwood and turf designs over thousands of years. Overall, this whole structure feels strong, but in a soft way, as a visitor once put it. As in a basket, there's a bit of give in everything; it's not completely rigid.

Inner Uprights

There are thirteen inner uprights, each supporting one of the main rafters. For the exact location of each, Jane and I walked around the inside of the structure we had erected, and imagined how much space would be required for each sector. As we felt OK about the space for that, we put a stick in the ground and moved on round. We went round clockwise talking it through: "Here we are in the coats area; we come in the door, kick off our boots, hang a wet coat on the post... here. Now we are in the clothes storage area - two racks of clothes hang from the wall to a low partition... here." etc. We wanted an inner room big enough to dance in, and have band practices or meetings, so we tried to keep the outer ring reasonably narrow. In practice this worked out at the inner circle being about 2m in from the outer one, widening a bit in the north and narrowing a bit in the south sun bit/bath area. I've never actually measured it all the way round. Hard to believe, isn't it, but I didn't need to.

The way we built the inner henge was to put each upright in first, then fit the crosspieces in situ. Kit form would have been possible, but this way each piece is custom made to its own rafters - less chance for the working of the great Murphy's Law ("Yea, if something can go wrong, it will"). Going round to each rafter in turn with a ladder, I moved a plumb line up or down the rafter until the lead was at its closest to our stick marks. Jane then put a new marker stick to mark the place exactly; I marked the rafter where the new upright would meet it, measured how long the support would be, marking it down on a piece of paper, and we moved clockwise to the next. When we had finished, Jane started digging the holes a cubit deep, and I started dragging the poles into position, their bottom ends already debarked and creosoted. These poles are the heaviest in the structure, because of their length, and they are all at least a hand span thick, so each one must weigh at least 100 kilos. I cut them to length with an angled saw cut (*diagram 6*) to give a better fit at the rafter, and also debarked and flattened the rafter at this point. If you want to be ethnic and use a good scrabble word here, pick up your adze. If you don't have an adze, you can do this kind of job quite adequately with a hand axe tapped like a chisel with a hammer or mallet.

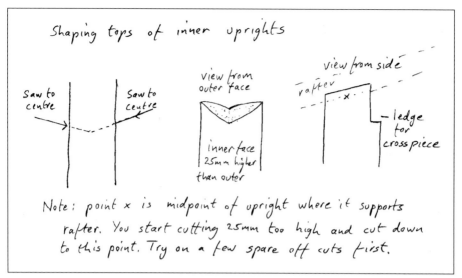

Shaping tops of inner uprights

Saw to centre — Saw to centre

View from outer face

inner face 25mm higher than outer

View from side — rafter — x — ledge for cross piece

Note: point x is midpoint of upright where it supports rafter. You start cutting 25mm too high and cut down to this point. Try on a few spare off cuts first.

(diagram 6) Shaping detail for inner uprights.

When all was ready, and stones placed in the bottoms of the holes, three of us put each pole in its hole and raised it up from the centre of the circle to take advantage of the rafter's slope. (*diagram 7*) When the pole was touching the rafter, but still at a slight angle, I went up the ladder and thumped the pole into place with a heavy mallet until the pole was plumb vertical. If the pole is 1 or 2 cm too long, there is enough play in the long rafter to accommodate it. If the pole is too short, put stones in the hole until it is a tiny bit too long. In our case, miraculously, the poles were all just right, and after they had been thumped very satisfyingly to the vertical I fastened them with fixing strap to the rafters. Don't make this a final fixing - just a temporary fixing on the sides (although hammer 2 or 3 nails well in on the top of the rafter) so that the cross pieces can later use the same fixing strap.

Thumping the inner upright into place

mark

plumb

(diagram 7) Installing inner uprights.

29

Crosspieces

Crosspieces are only about 1m - 2m long, so will not be too heavy in hardwood. I used a few strong looking pieces of douglas fir, then pieces about upper arm thickness of ash, hazel and thornwood. (Any wood known for its strength would do. Oak would be good.) For each piece, I measured the distance between uprights where they met the rafters, and cut an inset into each upright to make a simple joint (*diagram 8*). It is quite a tricky one, because of the angles in three planes, and I got noticeably better at it as I went through all thirteen. You have to seat the left side of the cross-piece as firmly as possible, resting on the top of the upright. Flattening the side of the rafter at this point allows 25-30 mm for a ledge for it to grip. (The upright is thicker than the rafter, so there is a bit to spare.) In places, I got the cross-piece to overlap or rest on its neighbour, but this depended on their size, and the overall 3D configuration. It is obviously critical not to weaken the rafter here. I pre-drilled nail holes, then thumped the cross-pieces home with a mallet and used two longish threaded nails for a good grip. Finally the fixing straps are wrapped over the cross pieces and fastened down, linking as many pieces as possible.

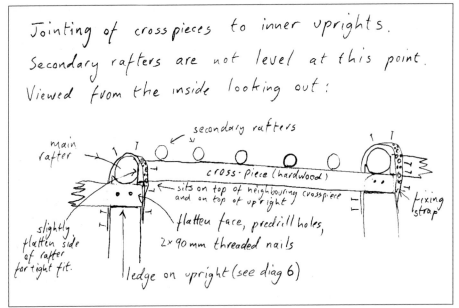

(diagram 8) Cross piece, inner upright and secondary rafter fixings.

Bracing

I have found the advice of other people invaluable at certain points in the construction of this roundhouse. Something like this really gets their imagination going, and I had some really useful tips from engineers, builders, architects and craftspeople. The bracing is thanks to a neighbour, John Hargreaves of St. Dogmaels, who pointed out to me that although the structure by now felt pretty solid, vertical poles with crosspieces could still all twist sideways under the stress of, say, a tree falling on the house. Several other people said no, it's fine as it is, but I could see his point. Diagonal bracing makes a huge difference to the rigidity of uprights, and it is better to be safe than sorry. So the six back walls have a big thick diagonal going from corner to corner, forming a kind of herring bone pattern around

Diagonal bracing on the rear arc.

half the circle. Each diagonal was offered to the space, and cut to fit into a notch cut into the uprights, thumped home by mallet, and fixed with 6" nails. It didn't take too long to do this - maybe a day - but after this the structure felt solid as a rock. The drawback is that it is much more difficult to infill a triangular space than a square space, so these wall sectors were fiddly when it came to filling in the walls.

The structure was now well defined, so felt good to be in and look out from. As I have mentioned earlier, we have on this community a simpler version, without the inner circle and bracing, that we use for gatherings and summer dances, parties, weddings etc. Brent made a grand canvas roof that can be suspended from the rafters and tied to the uprights, and we have canvas sides that can be laced together. A circle is the most efficient shape in terms of the space enclosed in relation to the structure material used (Rob Roy, in his book, *Cordwood Masonry Housebuilding*, works out that it is 43% more efficient than a rectangle, for example), so this circular frame construction is the best for the Tardis effect. You can fit a lot more in than you would think, and the space looks smaller on the outside than the inside.

31

The roof framework, before fitting inner crosspieces.

The Roof

Here is how the Miwok, natives of what is now northern California, built their assembly house roof:

"A thatch of brush, topped with digger or western yellow pine needles, never sugar pine needles, was next put on. This was followed by the final covering of earth. Altogether the roof was 1.5 or 2 feet thick. The opening in the top of the conical roof served as the smoke hole, the fire being built directly under it. The entrance was on any side. Certain niceties appear in placing brush and earth on the roof. The first layer of brush, which was laid radially over the numerous horizontal roof timbers, was of willow. On this another layer at right angles was placed. The third layer was of a shrub with many close parallel twigs that kept the earth covering from leaking through and resisted rot. The proper depth of the earth layer was 4 or 5 inches and was measured by thrusting in the hand. The proper depth came to the base of the thumb." (*Shelter*, quoting from The Californian Indians).

As far as I know, this roof is unique, and so far it has worked really well in keeping in heat, in letting in light, in providing habitats and growing places for countless plants and creatures, and in looking great from inside and outside. Working in Aluna after reading about roofs, it occurred to me to use the basic materials that the Miwok used, but to have a single waterproof membrane instead of the waterproof shrubs, and to use straw bales as insulation instead of pine needles. The straw bales could be tied together, so adding strength as a kind of skep sitting on the rafters. (A skep is not only another great scrabble word but the old type of beehive woven from straw.) For the membrane I phoned up several pond liner makers for samples. PVC is cheaper than rubber, but its manufacture is grossly polluting. (Greenpeace did well to persuade the government not to use PVC for the Millenium Dome for just this reason.) Rubber extraction is one of the few ways that native dwellers in the tropical rainforests can make a living sustainably, while preserving the balance of the forest, so we determined to use rubber for the membrane. This is the single most expensive item in the building, costing about £650 for a 40ft square/147sq.m sheet of 1mm thick rubber with a guarantee of 30 years' life. When it arrived (from Stephens Plastics, Wiltshire, tel.01225 810 324/5/6), it was in a huge bundle weighing over 300 kilos - nearly a third of a ton! We would need some help manipulating this.

The roof design incorporates two kinds of labour input. The first is the usual one of me, Jane and maybe another helper pootling away at some long task that can nevertheless be done by one person. This would all be in preparation for the second mode - one dry day of ant-like activity by maybe 20 people, carrying and humping up bales of straw, tying them together, and then covering the whole thing with the enormous pond liner. After that, with the structure effectively under a giant umbrella, we could return to mode one, and carry on building the walls and turfing the roof in ones and twos. (This is actually a main reason for choosing a 'post and beam' way of building, which this is. You may remember that such a system, i.e. roof before walls, was proposed by Noddy in *Noddy Builds a House*, by Enid Blyton, but laughed out of court by Big Ears. The advantage is that, after making your skeleton, you can put your roof on and then work in the dry. Makes a big difference, especially with a Welsh climate. Post and beam also offers a better resistance to earthquake damage.)

Jane, Julian and I are members of a local permaculture group that for several years met on the third Sunday of each month at different people's houses to carry out some work, usually of a heavy gardening/landscaping nature, that would have been very onerous for that person to do individually. Over the years we cleared a lot of bracken and brambles, planted thousands of trees, and dug several fine ponds. We booked our group to come here on Sunday, Oct 19th 1997, and prayed for sun. Here is the roof in cross-section (*diagram 9*).

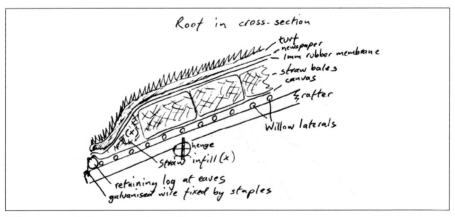

(diagram 9) Cross section of roof at eaves.

Secondary Rafters

For each rafter there are seven or eight secondary rafters parallel with it and spaced evenly about a hand span apart, resting with their light ends on the next main rafter, and with their heavy end resting on the outer henge. Secondary rafters therefore vary in size from 6m or so (the one nearest the main rafter) to approx 2m nearest the next rafter (*diagram 10*). Putting these up was easy. I was glad that we had thinned so many trees, and therefore had a good choice of over 100 poles to use. I fixed both ends with a double loop of telephone wire, stapled to the pole and to the henge or rafter. Telephone wire is great if you can get it, because it's designed to withstand enormous tension over the distance between telegraph poles. I was given about a quarter of a mile of the stuff by BT engineers after the January storms of 1990 brought down most of the telephone lines near us, and have been using it on structures ever since. Heaven knows what I'd have used otherwise. Maybe I'd have taken up stone masonry instead. Anyway, if you are building something like this you're going to have to exercise your imagination as to what to use. Fixing strap is fine, if ugly, but would be very expensive in the quantities needed. (At least 100 metres) Old electric fencing isn't bad, and old wiring is OK. Look in skips outside banks being refurbished. I found about 20m of telephone wire being thrown away outside Lloyds Bank in Cardigan. Windows too. It's amazing what banks chuck out.

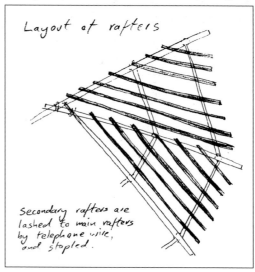

When all the secondary rafters were on, we could see the pattern of the roof underneath for the first time. I'd only ever seen it before in Aluna. I'm very pleased with it - it's like a giant iris of an eye. It became clear at this point that the internal design should not impede the clean sweep of the lines of the rafters, so that meant new thoughts about partitions, etc.

(diagram 10) Secondary rafter layout.

Willow Laterals

The next task was to weave and fix on the next lateral layer, which is mainly willow wands and saplings of about thumb thickness. These were cut green, all from within 100m of the house, and used immediately. The best withies were from a coppiced willow, or where a large willow tree has fallen over but continued growing, as they always do, by sending up dozens of shoots along the now horizontal trunk. Shoots of coppiced hazel are good, too.

The aim of these two layers, of course, is to spread the weight of the turf as evenly as possible over the entire roof, so I wanted lengths of willow no more than a hand span apart over the entire roof. In some cases we managed to weave the strands in between the rafters, but we found that tying with string or nailing was more practical. This task took a lot of time to finish, and was a rather uncomfortable working position, although the view was fabulous.

Jane willowing the roof.

A light coloured tarpaulin covers the willowed framework.

Materials

Before the Permaculture Roof Day I had to make sure that everything necessary would be to hand, so we brought down the membrane on the tractor trailer, amassed a fortune in lengths of baler twine, and bought and collected by Land Rover and trailer 150 bales of straw. These cost only £100. Straw must surely be the most under-rated insulating and building material in Britain. Another important layer in the roof is the canvas, which was to be tied onto the willow layer. Its purpose is to prevent bits of straw falling down into the house, and generally to contain the straw layer. I asked Willow for any information he had on old tarpaulins that might be suitable, and he found for us two small tarps and one giant white piece of canvas that had once been the side of a circus marquee. The whiteness was good, too, as a light ceiling helps brighten a room.

Roof Raising

The big day dawned unseasonally warm and sunny. Twenty friends and relatives rolled up, with food to share for lunch. Julian and two twelve year olds started wheelbarrowing straw bales from the stack to the eaves. Philippa formed a team tying on the canvasses.

When the canvas was on, Willow and Emma formed a chain heaving the bales onto the roof, and others joined them. I started with a square of bales round the roof hole, and worked out in a spiral. Keith tied each new bale to the one behind, and I tied it to the one or two above. Children helped stuff any gaps with loose straw, and got into a whole social scene on top of the house as the giant skep grew in size and the day progressed. Baler twine became as rare as gold dust, but just lasted out. We stopped about 1m from the edge, to allow for a slope down to the eaves (no point putting 30cm of insulation on the roof outside the outer wall, is there?) There were 30 bales left, so we had used 120. That's a lot of insulation.

Meanwhile a team of sappers had been constructing a wooden bridge from the bank behind the house onto the roof. They then unfolded the huge membrane and rolled it up into a 40 ft long sausage. As the light was fading and Keith crotcheted his last bale link, we all lifted a bit of the sausage and marched over the bridge in a long line onto the roof, put it down and unrolled it. The straw was safe, the cover was on, (and the roof could hold twenty people). We went for a swim in the stream to wash the dust off. I think this was one of the best days of my life.

Tying bales.

38

A team of helpers move the 120 bales into position on the roof.

Pulling on the rubber membrane.

Measuring up for the eave logs.

The policy we adopted from this point was to work on the eaves and, subsequently, turfing, when the weather was dry, and to do the cordwood walls when it was raining. Here is what we did to finish the roof.

First of all I went round with a sharp knife and cut off the four corners hanging down, and trimmed the whole sheet to about a foot (30cm) all round. The spare rubber was useful in dozens of ways - primarily to put at the base of each wall as a moisture barrier.

A visiting architect recently looked at the rainwater dripping round our eaves at various points and sighed 'Oh, if only I could get away with details like that!'. What he meant, in the nicest possible way, was that these eaves are some of the most unprofessional, untogether, and, well, organic features of the whole place. I explained that, rather than try to channel the water running off the eaves into a system of gutters, which would be a means of saving rainwater and sparing the rafter ends from rot, I had simply put a system of turf-retaining logs that would be easy to replace when they rotted away. This is because the pattern of the rafters coming out over the henge is really irregular. It would have been possible to make a system of run-off points around the circumference where water spouted off in a predictable fashion, but I really like the non-linear feature of it just as it is. If we did not have a free source of excellent, clean water from the mountain, I would also have put more energy into water conservation, as I have done on structures in the past. Jane and I lived in a bungalow served by a water meter before moving here, so were able to see that by employing water conservation measures you can save over two-thirds of your water consumption. In this case, however, we have an eaves solution that is dead cheap and looks very natural. After all, one of my personal mottos is *semper ut funquet*, let it always be funky. (The other one is *Why mess about?*)

We have, therefore, a system of eaves that has as its main aim the retaining of turves on the roof, and protecting the edges of the outer turves from drying out too quickly. It also has to prevent water reaching the outer walls of the house, and to prevent water from soaking the straw insulation layer in the roof. Even without a gutter system, it was still quite fiddly to do, and took several days.

The first task was to extend, where necessary, and make good the canvas layer on top of the willow laterals, and to contain the straw layer adequately. I used several ex-army blankets to achieve this. In some places there are small gaps where a small bird or bat could get into the straw under the eaves and make a nice cosy nest there. That's OK by us. We humans have deprived so many of our fellow creatures of their natural habitats that the least we can do is to put new habitat opportunities into every thing we build. You will see from *diagram 9* that this under layer meets the rubber at the outer edge of the eaves. The two layers are folded together and pinned at about 20cm intervals to the edging logs, which are linked loosely to each other and to the rafters by means of short lengths of galvanised wire and staples. This might seem a very rough and ready system to you, dear reader, but we are talking very funky 3D here, and it works. The factor that I have so far not mentioned in this design is the slope of the turf roof at the edge. We have placed straw bales up to 1m from the edge. These bales are one foot (30cm) thick, and the roof is sloping. The challenge is to fill the triangular space ('x' on *diagram 9, page 34*) with straw to give

Eaves detail at rear; note the angle of slope and random edging logs.

the most gentle slope possible so that the turf will stay on. In practice I found the best way of doing this was to combine all the folding, stapling, cutting and straw-stuffing in short stages of about two metres at a time, working round the perimeter. I measured the next length where one log could comfortably touch a number of rafters, from the first to the last rafter. This might be anything from 2 metres down to a tricky bit where a short length of only a foot (30cm) would be needed. I cut a log to size and offered it to the rafters to check the fit. I then cut four or five handspan lengths of wire, and stapled the wire one piece to each end of the log, and two or three to parts of the log that came nearest to the rafters underneath. Then I stapled the wires to the rafters, and one wire to the last edging log. Then I gathered together the canvas with the rubber membrane, folded them together, and pinned them at intervals to the log, just over the top of the log if possible to delay the action of water making its way through the nail hole to rot the log. Am I boring you? OK let us proceed. I was then able to stuff the space 'x' thus created with new straw taken from the 30 unused bales, and this procedure, bit by bit round the house, used up 20 of them.

One thing I dislike about modern buildings is all the straight edges that are so unnatural. I wanted an edge to the roof that had texture and wavy edges, and I am very happy with the look of the finished result. I reckon that the edging logs used here (mostly offcuts from secondary rafters, approx 4"/100mm diameter) will be the first things to rot in this house, but so far none have. When I am out cutting firewood and I find a log covered in moss, I bring the moss back and stuff it on or around one of these edging logs. So maybe, if the moss takes and builds up a good layer, I won't need to replace the logs when they go, although that would be easy to do. Rob Roy, the American cordwood masonry builder and writer, started edging turf roofs with huge ex-railway sleepers. His latest buildings are edged with clumps of moss. Much easier on the back. I am also hoping that the variety of fruiting and other plants in the roof will in time build up a growing edge that transforms the drudge of building maintenance into the much more pleasant gardening tasks of weeding and pruning. The roof was now ready for turfin'.

Turfin'

I'm sorry, I loved the Beach Boys. I can't help it; every time I start talking about turfing I start singing surfin' songs and making awful puns. Compassion, dear reader; that's all I ask.

Why turf, rather than, say, shingles or tiles or slates? Well, for one thing, I honestly believe that if our ancestors had invented a way of making single large waterproof sheets that were of organic materials and lasted 30 years or more, the whole look of our civilisation would have been different. Most modular roof systems, like slates, were a very time consuming, heavy and expensive way of trying to keep water out over a reasonable surface area, and needed a steep, flat slope with strong inflexible supports to hold them up. Hence our ideas of vistas of pointed roofs, gable ends, and all the paraphernalia that goes with what we think a group of 'houses' looks like. We are almost stuck with this shape, based though it is on technological limitations of the middle ages. (Planners will usually only give permission for a 'house' with all the cultural and obsolete baggage that the idea of a house entails. Nowadays, of course, a 'house' usually has a 'garage' next to it, or built into it. Will planners of the 23rd century still be requiring country 'houses' to have a traditional garage attached to them, even though cars are long obsolete?) The most organic system of our ancestors for keeping the rain out was thatch, which looks good, provides habitat, is renewable, and is warm if you can get it in sufficient quantities, but only lasts a decade or so before it needs replacing. Imagine if rubber pond liners had been available 200 years ago. Our expectation of what a human settlement looks like might be more like a badger sett with skylights. Technology doesn't have to be big and hard and shiny and fighting nature. Technology could lead us back to a more harmonious relationship with nature, and if this house demonstrates only that, its construction will have been worthwhile.

For me the essence of a turf roof is the fact that I haven't taken any surface of the earth away from nature - I've just raised it up a bit. Plants and creatures can still live there. In fact, since it isn't grazed or prone to bracken invasion, it provides a new, protected habitat. The other reason to go for turf is for its low visual impact, which for me is not an intellectual concept or a planning buzzword - it's just natural. How many other animals, except humans, can you think of that build their dens on the tops of hills and in other places that stick out for miles and announce their presence to the whole region? The more I live in a natural secluded den the more I see these loud, rectangular, mock-permanent human living zones as a severe aberration. More about this later, maybe, but for now let us just say that for me it has to be turf. With a few tayberries and strawberries in it, of course.

Tayberries on the roof edge.

There are some good books and leaflets on turf roofs (*see appendix*). Some people prefer to chuck loads of earth on and then seed it. We prefer to cut individual turves from nearby, load up a wheelbarrow, hand or carry them up to the roof and lay them then and there. In the long run it doesn't make much difference. The celebrated architect and turf roof maker, Christopher Day, who lives just over the hill, once told me that the turf roof on the Steiner Kindergarten school at Llanycefn, made in the early 1980s, has been found to have 23 species of grass, none of which was there in the first place! So a turf roof is a living thing which evolves to suit the conditions in which it finds itself.

There are a few factors to consider in turfin'. Slope, holes, roots and water. The rest is just hard work with a good view.

When grass roots grow down and hit a waterproof membrane, they grow sideways and mesh together in a strong web. At first, however, for the first few weeks, each turf is on its own. A new turf will slip off a smooth surface if a lot of rain saturates it and the root system is still inadequate to bind the whole section together. So getting a healthy turf to stay on is a question of grip and root system, rather than a simple slope formula. Some texts will tell you that a turf roof must not exceed a 10%, or 15%, slope. Although you will have fewer problems with a gentle slope like this, there are cliff edges on the Pembrokeshire coast a couple of miles from here where grasses and wild flowers are clinging on at angles of over 45%, in defiance of all the textbooks.

For a roof shape, I personally prefer the 'mushroom' kind of look where most of the roof has very thick insulation that comes down to nothing at the eaves, thereby giving a sharper incline at the edges. The way Jane and I taught ourselves to apply turves allows us to lay them on quite a steep slope at the edges. We start at the edge, and work towards the centre. The edging turves are twice as long as the more central turves. For a normal turf, we have found about a square foot (30cm x 30cm) area, and about four fingers depth of soil (like the Miwok) is about right. If you are building something with seriously strong roof members, say oak, or pine beams 10" x 3" (25 x 75cm), then you can use considerably thicker turves. This will give you slightly improved insulation (but not much - soil is not brilliant as insulation), and greater potential for drought resistance and deeper rooting plants. We don't send back a turf if it is 6" deep (18cm), but a series of them plays heavily on your arm and back muscles. An edging turf is twice as long but the same width and thickness, and we use these for the first two rows at the edge. One person can just manage to carry one of these, and you will get about six in a wheelbarrow. For cutting, use a sharp spade (yes, you can sharpen a spade on a grindstone. It makes a huge difference. Mind your toes). When you have a large number of turves to cut, i.e. approx. 1400 turves on this

Turfer girls Martha and Jane; note the long turves on the bottom edge.

roof, it saves a lot of time to make a standard size and to be fairly methodical in the cutting. Jigsaw puzzles are fun, but not on this scale. So we plan out the area, and cut in a sequence.

Before laying the turf, it is a good idea to place a newspaper or some absorbent fibrous material such as old blankets on the rubber. You can buy 'geotextiles' for this purpose. This serves the purposes of protecting the rubber from small pointed stones or thorns that might pierce it when a person walks above, and of providing a medium for roots to form a mat in. On some parts of this roof we did not put this newspaper lining (several pages thick), and at times the next summer the turf became very dry. So I recommend that you do it all over.

Just a note on holes. Make sure not to allow a single hole to go unrepaired before turfing. One day while I was turfing a whole lot of nails spewed from my pocket onto the rubber and I trod two into it before I noticed. 18 months later we noticed two damp patches in the canvas ceiling, and needed to spend a morning chasing up and puncture-repairing those two holes!

Laying the turves is excellent exercise. We started where we could reach, to get into practice. The edging turves are laid lengthwise up the slope, so that their root structure holds them on the steepest bit. I make sure that the bottom end of the turf sits firmly in the depression behind the edging log, and thump it into place with my fist. As we moved round the eaves to the south part which is about 2.5metres above the ground, it was a matter of climbing a step ladder with a long edging turf, cradled in such a way that I would fall into place with it, aiming at the exact spot. Any turf which split down the join was used as two normal turves for the next or third row. After the first two rows were complete the roof started to look, especially by moonlight, as I had imagined it. I can remember clearly the pure thrill of seeing this organic dream taking shape and becoming real. For the rest of the turves it was a matter of patience, arm muscles and good helpers. Most people in the community helped with a few hours of turfin'. Here's Martha helping Jane. For the normal turves, we devised a sling of rubber that would reach down to the ground and which held two or three turves. An efficient way of operating is to have one person digging, one person transporting and another laying the turves on the roof. By the end of January 1998 we got to the top circle, where the rubber was still uncut, stretched over the roof hole like a trampoline. Turf's Up. On to the skylight.

The Skylight

The skylight does not winch up, to let out smoke from a central fire, despite several people's romantic suggestions, although we did try it for a couple of days. Too much smoke stayed in. As a vent in hot climates this would be quite feasible, though. In Wales, however, keeping cool has proved to be fairly easy. The main function of the skylight is as a fabulous source of daylight, beaming down into the centre of the house in all weathers. I am typing by its light now, with a terrible wet and windy January day outside, and with no artificial lighting.

In composition the skylight is two slightly convex coach windows, each measuring 40 x 80 inches (approx 100 x 200 cms) and 6mm thick, laid one on top of the other with a rubber seal 2mm thick, sealed with silicon, between them. It's the best value double glazed skylight I can imagine. (The rubber, of course, came from offcuts.) The bottom window is laid over the hole, which measures 34" (86cm) in diameter, straight onto the grass. Turves are piled under the ends and on top of the top sheet to hold it down and to eliminate any tendency for the wind to get under the glass and make off with it.

Skylight from above, with turves being placed around the edges.

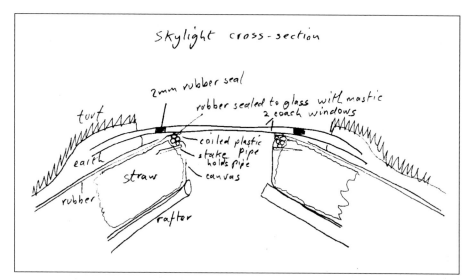

(diagram 11) Cross section of skylight.

Underneath, offcuts of canvas are tucked over and under the exposed straw bales to tidy the hole up. At the top of the hole, where the rubber membrane is close to the underside of the glass, I wrapped a piece of spare tubing around the straw circle a few times, supported on wooden pegs driven into the straw, to raise the rubber 1" (25mm) or so to touch the glass, then sealed the join all round with silicon sealant (*diagram 11*). Actually, there is still a slight air gap of a few mm on one side, which we don't mind as it allows for a bit of fresh air to make its way in up there. When there is a gale the rubber loosely touching the glass occasionally treats us to a sound best described as a giant raspberry.

The only thing to add to this is that we were especially careful to see that the sheets of glass were immaculately clean on both sides before we laid them on the roof and sealed them together, and that we did not get any sealant on the bit we wanted to be completely clear. Easier said than done - coach windows are heavy things. Just a tip on getting hold of big windscreens or toughened glass windows like this, if you are likely to want to: it's easier than you would think. I got seven of these; four for £5 each and three for free. On a new coach, each window is toughened, convex, thick glass of high optical quality. They cost over £250 each. When the coach is eventually scrapped, the windows will probably be in as good condition as when they were new, but obsolete. Unless some crazy permaculturalist with nothing better to do than build funky houses and greenhouses comes along, they are virtually useless. But few depot managers are willing to chuck out such beautiful artifacts as these that cost so much and must, surely, have some value. So I have, on several occasions, located algae-covered stashes of lovely windows and windscreens in bramble patches behind junk yards and bus depots after detailed cross-examination of the people working there. Don't ask the boss first; ask the guy maintaining the vehicles, then offer the boss money. No more than a fiver each, though - we don't want inflation to set in to the used windscreen market, do we?

The skylight lets an amazing amount of light into the interior.

Walls

*T*he walls of this roundhouse are mainly made of what is called cordwood masonry. It is, apparently, the first house in Britain constructed in this manner. The claim for first building of cordwood goes, as far as I know, to Ben Law, the Sussex permaculture designer, for his cordwood and turf workshop. (Ben and I have something else in common, which is that we both have almost identical African style wooden chairs, made by each of us on a permaculture camp from two halves of one beautiful oak plank.)

There are no inner walls, so this chapter refers to the outer walls. For the sake of clarity, I will call 'a wall' that section of wall between two uprights, although around 6/13 of the perimeter, from the West through the North to the East, it is really one stretch of curved wall, and most of the rest is window.

I started without the benefit of having read more than about two pages of text on cordwood masonry. In Ken Kern's book *The Owner Built Home* there is a fuzzy photo of an old guy in the Mid-West beside a wall made of log ends that he was covering with chicken wire before rendering. I have been wanting to try out this method for over twenty years, since seeing this photo. *The Short Log and Timber Building Book* by James Mitchell gave one useful clue: to put a layer of insulation in between the outer and inner mortar layers. Everything else I made up, but judging by a read of Rob Roy's book *Complete Book of Cordwood Masonry Housebuilding* (which I have subsequently obtained, in exchange for a wooden bowl, from the author when he paid us a most enjoyable visit), I hit upon reasonably acceptable ways of doing it. The main difference from the straight American system is that we had resolved to use no cement in the construction of this house, so in place of Rob's mortar of cement, sand and damp sawdust, we used raw mud from two paces outside the eaves - the big pile of subsoil that came out of the bank that this den is set into. The earth was high in clay content, with some sand and grit; in fact, as far as I can see it merits the term 'cob'. We mixed it by the barrow load with water from buckets placed under the eaves (it rained almost every day) and three or four handfulls of shredded straw or dead bracken. The insulation is straw. Rob recommends sawdust, which would have done equally well, maybe better,

51

but straw works fine. The log-ends are all as near as I could get to 16"/40cm without measuring. I cut maybe two hundred at a time with the electric chainsaw up by my workshop and brought them down by the tractor and trailer. It is good to have a big pile of logs to choose from, so that you can keep ringing the changes in size and texture. There should always be the perfect log-end waiting to be laid in the next space. If it's not there, you need to cut another load.

I started with the wall in the North-West; the one next to the front door/window. This would mean we could work our way clockwise (sunwise for New Age readers) all round the perimeter. The foundation of each wall is the same - a raised platform of puddled cob maybe 4"/10cm high and a bit wider than 40cms, curving very slightly outwards and back to meet the next upright. On this foundation I laid one unbroken sheet of rubber covering the platform down to the ground at both sides and going up the upright a little at each end. This is to provide a damp proof course to prevent damp rising up the wall. Scoff away, O professional builder, but it works fine. We then prepared for wall laying, with Jane making up over three-quarters of the cob mixes. On to the rubber I laid a line maybe 5"/13cm wide of cob mix about two fingers deep, then an equal width of straw, then an equal width of cob again. This cob/straw sandwich method is used throughout the house.

Start of wall 2, showing curve of logs and damp proof course.

From one upright to the next is about two metres, so this is the length of wall I worked to, and it was an easy length to use in terms of support and handle-ability. What I mean by that is that the cob starts to thicken up and harden a bit after maybe half an hour after laying on, so that you don't want to lay cob down for more than a half hour or so's work. Two metre runs are about right for this.

Big log-ends on the bottom row.

On a good day I could get up to window height by lunch time.

I started off using a trowel to apply the cob mix, by the way, but quickly graduated to doing it all with hands in strong rubber gloves. Laying the logs was a piece of cake. Honestly, it comes so naturally that I could recommend it to anyone. The only explanation I can think of is that we have evolved over hundreds of thousands of years working with logs and mud. It's in our genes. I made the bottom couple of rows of good thick hefty log-ends, and if a log-end had a discernably thicker end, with this end to the outside. This gives the wall sections a slight curve. After the first two rows I went for a more random effect, choosing log-ends of all sizes and therefore using up all the wood at a standard rate. As I completed one row, or maybe a half row, I would place the mud and straw on top, select logs to fit, and tap them home with a lump hammer. The technique is slightly different if you are making a load-bearing wall with mortar between the log-ends; you need to allow more space between log-ends, as the mortar does most of the load bearing work. Here, however, I reckoned that what we were going for was the most thermally efficient way of plugging the space between uprights, which are bearing the load. Wood is better insulation than cob/straw/cob in these proportions, so I went for a maximum of wood and as little filler as I could get away with.

Pointing is finishing and making the cob smooth between log-ends. I pointed with fingers in gloves every three rows or so. Some people would say we left it too rough, and should have brushed the mud off before it dried. They could be right, but I like it how it is; it's a matter of taste.

Wall detail: cob, straw, cob infill.

The only tricky bit on these back walls came with the diagonal bracing, as I have mentioned. It is tricky because the diagonals are only 6"/15cm or so thick, yet the wall needs to be 16"/40 cms thick. At each upright and diagonal, therefore, I added shorter log-ends to make up the thickness. These log-ends need to be just as varied in thickness as the other log-ends, so as to maintain the same level when you clear the diagonal, so actually this involved having a large pile of 9"/23cm or so long logs for this diagonal infilling. I also used stones in a few places, and found that wine bottles were good in the small triangles at the top of the diagonals. After a while I got more practiced at using wine bottles for decorative purposes where the sun would hit them - put the bottle neck into a jam jar and use with the jam jar on the outside.

A similar problem to diagonals occurs when you get to the henge piece at the top of a wall. There is no space for big tasty logs, so you make do with squeezing in logs to fit. It's obvious, really, when you get to it. I left each wall like that, and came back later to tidy the walls up, sometimes with several layers of half length log-ends to mask the henge itself, which brought the wall up to the gaps between

Bottles used as fillers in window pier.

rafters - what the Americans call 'snow-blocking' (because here, if you left a gap, the snow would be sure to blow in, along with permanent draughts). What I chose to do was to fill these gaps with as many logs as possible, usually of a length a bit less than 40cm/16", and then to fill the remaining spaces with special hay balls. After a few experiments, I found the best way of making these was to lay maybe a quarter of a hay bale in a barrow of liquid cob and mix it a bit until the hanks of hay were saturated. Then I rolled a hank into a ball as big as a large snowball and stuffed it into the space. (I got the idea by seeing some brilliant insulation board made in Greece from thin hay-like seaweed and cement slurry - same principle.) After a week or so the balls had dried solid and have formed an excellent insulation and gap filler at the eaves.

Vents

On the subject of non-log filling, it is also worth thinking about any holes or vents that you want or might want in the future in a cordwood wall. Ken Kern suggests in his *Owner Built Home* that it is actually far easier to get adequate and targetted ventilation into a room by means of vents than by the usual method of opening windows. Sure, it's nice to be able to open the odd window, but all of them? Fitting a simple double glazing unit straight into a cordwood wall is easy and cheap.

Nearly complete; note large screwtop vent and rubber window flaps.

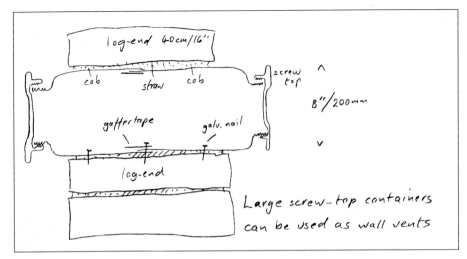

(diagram 12) Wall vent made from two large screwtop containers.

So, having planned for this, I also had to allow for plenty of vents to allow fresh air in and stale air out.

It is a very simple thing to substitute a piece of pipe into the wall at the building stage, so we have here about a dozen or more vents and pipes of different sizes varying from a piece of 1"/25mm tubing, through several pieces of 4"/100mm pipe, up to three giant utility holes made from two heavy duty plastic tubs with screw on lids fixed end to end, with their ends cut out. (*diagram 12*) These are quite fun to fit in the wall in place of a big log-end, and may even be of use one day for something or other. There is one in the NW wall where we keep a log pile near the back door, that could be opened to give extra ventilation for drying logs, or even to put logs through from outside, without the need to come in with an armfull of logs. In practice, the most useful vents have been the 4"/100mm ones over the cooking area, which are easily filled by a roll of carpet.

Cat Flap

Most cat flaps I have seen are made to fit in doors. If you live in an ordinary house with hard thick walls but thin wooden doors it makes sense to have cat flaps here, but it can be a real hassle falling over the cat on either side of the door when you come in or out. In addition, there's the competition and power games over the exit that humans keep getting in the way of, and also the little dead mouse gifts to beware of right in the doorway.

Better, really, when you're building your own place, to put a cat flap in the bottom of the wall, as near to or as far from the human door as you want. Ours is near the front door, on the other side of the front door upright, at the bottom of a small piece of wall filling one log wide. All I had to do was to support the first log-end at cat height with two side pieces of wood, and then continue up the wall as normal. To cover both ends of the 16"/40cm tunnel thus created, I tacked several strips of rubber from the tops of an old pair of wellington boots

The cat flap in use.

to act as a draught excluder. The cats needed to be shown once how it worked, then adopted it immediately. There is a vulnerable moment, as a cat, when you are passing through the tunnel which is as long as you are, with no chance of turning back; it gives hours of fun to kittens, and still a few nervous seconds to the adult cats, as would the overhanging rock opening to the small cave that their wild cousins choose.

Straw Bales

I don't want to say much about straw bales here except to mention that in several places in the walls - on straightforward stretches where no window was required and the wall would be above ground level - we used straw bales instead of cordwood. They are simple to use and very quick to lay; they are easy to plaster with a slightly more liquid cob mix, preferably with a bit of horse manure in there for plasticity, and they are very good insulation. You can make curved walls with straw bales that are structurally stronger than straight walls with rectangular corners. The only problem with them, as far as I can see, is that small bales are almost obsolete, and that most farmers now produce those huge round bales. In this area, where many farmers maintain old equipment with great reverence, however (our neighbour uses a baler, which we borrow, that looks like it was repainted in Carnaby Street, circa 1967), bales are still easy to come by.

Platforms and Partitions

One great advantage of cordwood walls filling a post and beam frame is that it is possible to use the upright posts as anchors for platforms, shelves, partitions or other features, and to build the wall right round them. For example, when we got to wall five, in the NNW, it was time to start thinking of the sleeping platform we wanted at about 1.5m/4'6" above ground level. I wanted this platform to cover two wall sections, so as to be long enough not only to provide a long space for a bed, but also 2m or so of platform on which to store our fairly extensive range of musical instruments. I bought a dozen strong steel bolts about 13mm thick and about 10"/25cm long, plus some lengths of threaded bar and bolts and washers. Brent offered to help me for a couple of days, as I wanted to use fairly heavy members for the bed platform, so we flattened the sides of inner and outer uprights at the same height, using a spirit level, and flattened the ends of the platform logs, which were basically similar round logs of Douglas fir as the framework; maybe a bit thinner, about 8"/20cm. We marked out where each support would go, and drilled a 13mm/.5" hole with a brace and bit through the log ends and the uprights. In some cases we could use a long bolt countersunk into the pieces, and in some we used the threaded bar, cut to size, and bolts.

Platform framework.

The end result was a sturdy grid of 5 horizontal radial members which would support a framework of smaller logs laid across them. The final platform of old floorboards lies on and is nailed to this framework. This system proved to be very simple to construct and is rock solid. Underneath is a large storage area for tents, sleeping bags, junk, wine and logs. When it came to laying the log-ends in the next section of wall, I simply laid up to and

Willowing the roof, with views to an ancient forest beyond.

A job well done, Tony sitting in front of the completed roundhouse.

The main living area with sleeping beyond and kitchen to right.

The sun bit, a relaxing spot with meadow view, hot wall at end.

Wine fermentors, plants and a cat share the bathroom windowsill.

This strawbale walled den is almost invisible in its woodland setting.

Jane takes time out in the den. Note the impressive windscreen window.

Ladder to sleeping platform.

around the members bolted to the uprights. It looks as though one of the log-ends just continues out of the wall to hold up the platform. The bolt is concealed by adjacent log-ends. The picture on left shows platform supports by the back door, also hazel ladder to bed, small window by bed, log pile underneath and screwcap vent in wall. I used this idea again for the platform holding the big hot water whisky barrel - again something needing structural strength

Windows and Doors

I had no idea how to fix windows into cordwood masonry when I did the first one in the W wall, but since there turned out to be 11 windows in total, I had got the hang of it by the time we got round to the West again. Only one window here opens, and that is the little one by the bed; we can have a fresh breeze in bed if we want it and hear the birds singing. That one, plus the one which is a tiny slit of light at the top of the N wall, and the giant coach windows are the only single glazed windows. The rest are double glazing units or double glazed patio doors.

Basic window design

As all the windows are different, I wanted to avoid the impression of total chaos by giving as much order to their appearance as possible. One way this is done is by using the smallest ones in the NW, N and NE and gradually using bigger and bigger expanses of glass until the S walls are floor to ceiling glass. (In case you're

A sealed double glazing unit, clearly showing the slight downward tilt.

thinking "Ooh, all that reflected sunlight flashing away into a natural spot", I set them all very slightly to face downwards by maybe 3% off the vertical, so that sunlight does not reflect off them horizontally.)

The other way of regularising them was to place window sills in a line at the same level to support the three windows in the E and SE, and to continue this level with a window divider on the next floor-to-ceiling window in the S. I think this layout has worked, both aesthetically and in terms of solar gain. Although you don't see that much glass from the outside, there is in fact a large expanse of glass where it counts. The eaves prevent too much summer sun coming in, but at the moment as I sit at 4.20pm on a February afternoon, the sun is streaming in through the W and SW windows, right across the central area and flue wall and shining on the pots and pans in the kitchen at the far side of the house. Everything is warm and brightly illuminated, and the view out from these windows is of a continuous panorama of hazel trees and their new catkins just to the east, the willows and gorse in the south, and the mountain showing through the leafless alder, oaks and ash in the west. From the 'sun bit' we can sit and watch woodpeckers raiding anthills a few yards away, and see the many species of small wild birds that make the woodland edge their home. For this is what windows are about, isn't it? Bringing nature into the home.

There is a limit to the usefulness of my giving you all the details of the particular windows here, because the chances of you personally getting an identical load from Kevin are pretty low. Better to mention any salient points that might be applicable generally. I have been quite happy with so few windows opening. It is much easier to fit a simple unit, and you don't have to bother about how and where the window will open. The opening window by the bed has a basic simple wooden box made of old floorboards around it. The box is stuck in the wall and log-ends put in all around them. The other 'back' windows, which are stuck fairly high in the wall from the inside view, but which are all at ground level on the outside, have a common pattern of hardwood sill placed on the built up log-ends, with a rubber sheet between them and the sill.

Rubber sill sheets

To be honest I don't exactly know why this sheet is there under each window, but it will no doubt be made clear to me one day! I think I had the idea that if driving rain hit the window, the water might stream off the window, soak the sill, and start to erode the cob between the log-ends below. I left a flap of any surplus rubber hanging down over the top rows of log-ends to protect them. In practice, this house is so sheltered from the N and E, and must have such a low wind drag factor that we have never had rain driving against these back windows, protected as they are by a half metre of eaves immediately above. I could, in fact, have left the rubber out, but it is still there, maybe because another of its functions is still valid - on the front of the house, on the more exposed side, the long rubber overhang acts as a good draught excluder and insulates the dry wall from the potentially wet, or at least seasoning, log pile stacked against it on the outside. I wouldn't leave the outside bare, i.e. with the rubber sheets visible, because I think it looks rather ugly This aesthetic consideration is another good reason, in fact, in addition to the obvious one of having a good pile of seasoning firewood, for keeping the log pile up to date and in prime condition. A visiting lecturer in architecture was admiring the aesthetic down-home look of the woodpile. 'Of course, you can't burn it', he said.

Lintels

But I digress. The basic window design evolved. Wood and cob both shrink when they dry, and I thought it better that at least the windows stay in the same place, even if the wall shrinks or moves a bit round them. I learned from bitter experience, with a greenhouse I built about 10 years ago, that glass itself does not shrink, so you have to allow for the wall to move if it wants to, without it trying to force the glass to move too, and thereby cracking it. A fairly simple way of fixing the window onto the sill, which has so far worked perfectly, touch wood, is to sit the unit on a rubber seal which is held in place by glue and tacks. I then put a few small galvanised nails into the sill on each side of the glass to hold it on the rubber, as glaziers do under the putty. At the top of each window, rather than one fixed lintel there is a double lintel to hold the weight of any log-ends above. The lintel is effectively in two halves, which gently

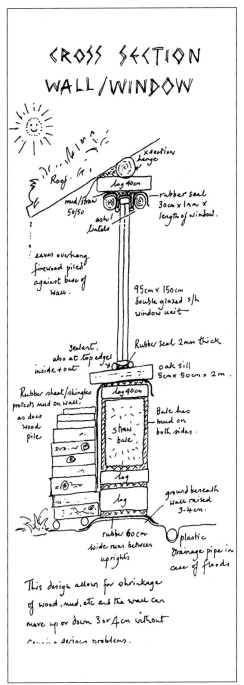

CROSS SECTION WALL/WINDOW

Roof.

x section henge

log 40 cm

rubber seal 30cm x 1mm x length of window.

mud/straw 50/50

ash/ Lintels

eaves overhang. firewood piled against base of wall.

95cm x 150cm double glazed s/h window unit

Sealant, also at top edges inside + out

Rubber seal 2mm thick

oak sill 5cm x 50cm x 2m

Rubber sheet/shingles protects mud on wall: as does wood pile

log 40cm

Bale has mud on both sides.

straw bale

log

log

ground beneath wall raised 3-4cm.

rubber 60cm wide runs between uprights

plastic Drainage pipe in case of floods

This design allows for shrinkage of wood, mud, etc and the wall can move up or down 3 or 4 cm without causing serious problems.

(diagram 13) **Cross section of wall.**

hold the unit on either side at the top. The window is sealed at the top as well by a rubber flange. (Yet another use for pond liner offcuts). In this way, the wall can settle a centimetre or two without the window taking any strain. The lintels simply rest on the log-ends, with the smaller windows. With the bigger windows I took the lintels right across and fixed them to the uprights, then built the log-ends up to and over them *(diagram 13)*. It's just common sense, really. If you ever do this kind of building, you'll probably make up your own way.

Back door

The first variation with this pattern came in the E with the back door, which is a double glazed double patio door. (Don't laugh - luxury is OK if it only costs you a tenner). The two doors of this unit fit inside a rectangular metal frame, so all I had to do here was to fit a floor plate, made of one long douglas fir log, onto the ground, raised at just the right height and exactly parallel with the henge piece above so that the whole door unit fitted snugly between them, with one upright forming one side and another long upright at the other side. I then screwed the metal frame to this log frame, first checking that the corners were square. (You do this by measuring from one corner to the one diagonally opposite. Then measure the other diagonal. If they measure the same, the corners are 90 degrees - 'square'.) Then put the glass doors in. It worked perfectly - the opening door slides across the fixed half with a gentle hiss.

By this point we had almost half the perimeter of the house with walls and windows in, plus a fancy patio door to impress visitors. The weather turned wintry. Temperatures outside, and inside our little bender before we got the fire going in the evenings, dropped below zero. Jane and I whacked some old floorboards onto the sleeping platform, shooed the chickens off our old bed and mattress stored in the farm stable, brought them down on the tractor (the bed things, not the chickens), and moved in. I rigged up a simple chortle, or milk churn stove, filled it with offcuts, and lit it. Our new environment was wild and muddy chaos, but at least we could stand up straight in it.

South windows

The next three walls of windows needed more work in Aluna, as there would be more window than wall. The picture I got was of a series of big sills, broad enough to extend outwards to cover the straw bales or log-ends, and inwards to give a good standing area for wine demijohns, pot plants etc. These sills manifested as 2"/50mm thick old oak boards over 2m/6'6" long that had been planked and stacked in a neat pile in the woods here for about ten years and were crying out for a new lease of life. I built the lower parts of the wall up with log-ends and straw bales, added the rubber sheets onto a bed of cob, then fixed each sill into horizontal slots cut into the uprights at each side. Thus the weight of these windows is supported by more than just the log-ends. (*diagram 14*). I sealed these three windows with silicon sealer at top and bottom, and with cob up the sides.

For the sides of a double-glazing unit, usually you can get away simply with filling up the space between the window and the uprights with log-ends and cob, siting the window half way along the logs. This is a bit tricky, because the edge of the glass unit is itself holding a whole lot of logs in place. Cob is very malleable stuff, however, and I found that it

Labels on the diagram:
lintel pinned to upright

75 mm pipe for ventilation

ash lintel

mud + log wall

sill fits into upright

oak sill

(diagram 14) Window lintel and sill fixing detail.

will set well and hold OK. The advantage of such a simple way of framing the window is that you can shape the corners at the cob/glass interface with pleasing curves. I did all the back windows this way. When it came to using very large units in the S, however, one on top of the other, it felt like they needed some kind of wooden frame all round. Also, if any windows were to get a real battering from the elements, it would be these ones in the S and SW. Rather than go back to the box method, which would look too rectangular for a high profile part of the house (*semper ut funquet*), I chose pieces of round, straight ash to grip and support the sides. Long enough pieces to fix to the ground plate and with spare to join to the centre divider or lintels at the top. I carried the ash pieces up to the workshop and with the electric chainsaw cut a groove wide enough to hold the units with 2mm to spare on both sides, and approx 1"/25mm deep. Obviously no wild piece of wood is going to be dead straight, so it is important that the bottom of the groove is straight, despite the odd divergence on the surface of the wood. I then brought the pieces back down again, put silicon seal in the groove, pre-drilled screw holes (hammer and nails don't go with glass, I find) and screwed this rough hardwood frame to the footplate and lintels. No problems with this so far. The main thing to watch out for is allowing enough spare in your groove for the ash, if it is still a bit green, to shrink a bit. That way it shrinks on to the seal, making it nicely watertight. I sealed these edges and frames with teak oil.

The most popular question people ask us about this den is 'how long did it take to build?' Well, on this bit, doing a wall section at a time, with wood log-ends, straw bales, glass bottles, sills, vents and windows in various combinations, it worked out almost

Mastic sealed windscreen window.

exactly at a wall a week. As each 'wall' is about 2m x 2m, it might be better to say that progress with this method was at the rate of 4sq. metres a week.

Front door

Front doors say a lot, don't they? Castle or burrow? Come in or go away? Haven or retreat? Old money, new money or hovel? The first thing people do when they buy an ex-council house is change the front door. We waited several weeks while we turfed, infilled the back etc. until it became clear what to use for our front door. I lined up three or four good contenders and leaned them up near the space for a while. In the meantime we nailed up several overlapping pieces of spare canvas and blanket against the wind and frost, so it felt very yurt-ish. In the end we settled on the thing that would let in most light - a very heavy, metal framed and wood covered double glazed door of reinforced glass that must have been in some big classy building once. It was too tall to fit between the ground log and the henge, so I hung it on big farm gate hinges screwed onto the SSW upright, and fitted another log for it to abut onto. I put a strip of rubber down its edge so that it swings silently up against the upright and stays there, shut. I made a simple wooden catch to hold it shut against strong winds. That's it, really.

It took a bit of time and cursing because I had to drill through the metal frame, using a simple hand drill, to fix the bolts of the hinge on, but now it's there, it's good and solid. We could, in theory, in the unlikely event of a long, scorching summer, lift it completely off its chunky hinges, and leave the door open, but it weighs over a hundredweight, so we haven't so far been moved to try.

Front door (left), patio windows (right) and cat flap in central pier.

It didn't take us too many weeks to realise that we needed a decent flue and, come to that, a decent stove with back boiler, so I'll describe to you briefly what we have ended up with, which is great. Our milk churn 'chortle' was fun, but I love to see the flames of a real fire, and never got the chortle to work well with an oven-glass door, no matter how hard I tried. So after a year we bought a new 'Villager' flat top wood burner, with back boiler. This and the flue brought the total cost of this house from £2,500 up to £3,000, in fact. Gives us lovely hot water.

The flue is 5"/127mm steel, leaves the stove at the back and goes at a slight angle 2m/6'6"upwards through the stone and cob to the wall, where it is joined to a half-metre length of double skin stainless steel flue. This bit is basically a well insulated sleeve, to make sure that no heat makes its way into the wall. There are some stones and cob around the flue at this point as well - no wood touches it. The flue is then connected to 3m/10ft of 5"/127mm stainless steel flue by means of fitted angles, so the last 2m are vertical. The flue is fixed clear of the eaves by galvanised wire hawsers, and is topped by a cowl to prevent the rain coming in. In designing your flue system it is worth imagining the worst that can happen, then try to prevent a fire resulting from that. Wood burning stoves, and their flues, can get very hot if you stoke them up with good dry wood, you open all the air intakes, the phone rings and someone needs a lift or something, and you rush out without shutting the fire down. It happened to me once in an early den - I just got back as the stove and flue were glowing red hot and the twig basket nearby was sending up the first wisps of smoke! Imagine the worst, and design it out.

The flue here goes out through a wall because I couldn't think of a safe, water- and weather-proof way of leading a hot flue through a sheet of canvas, a foot of straw, then a sheet of rubber. Also, the hot wall is wonderful, so we are happy with the system. It does, however, mean fairly regular cleaning of the flue - at least once a month. If you build something like this, make sure to have an access hatch outside, at the angle before it rises vertically.

The Villager stove; note the copper piping to the hot water barrel, twig basket, clothes and towels drying above the hot wall.

Water

All the time that we were in the bender, and for a few weeks in the house, we relied for water on a small spring about 10 metres down the bank below the house. It is probably of good enough quality to be used for drinking, and sometimes we drank it, but usually we carried drinking water down from the main house supply in containers. This was slow and heavy, but at least you learn to appreciate the value of water. The main house used to be on the mains, but before that, within living memory, it had shared a local supply from a small source in a brick built covered well 400 metres up Carningli, the mountain immediately to the west of Brithdir. When Emma and Julian first moved in, they were told by the water company that the mains was the only proper, legal water supply, but they ran a new pipe to this old source anyway. The water was, and is, beautiful pure spring water. Within a few months all the neighbours had tee'd off the new pipe, so this locality is now on mountain water again. We extended the pipe down to here, so this is our water too.

I don't know about you, but the subject of plumbing makes me drowsy even to think about. I hate crawling about under things trying to fit pipes together. I'd miss this bit out if I could, but there you go - mind over matter; the superior man perseveres, etc.

I tried thinking in a reasonably permacultural kind of way about our water system, and I am pleasantly surprised to say that it is simple but it works. Cold water comes into a header tank located at a height to give enough pressure to push the cold in at the bottom of the hot tank and out, heated, at the top. That is about half of my total grasp of plumbing. I didn't want to have too many ugly tanks spoiling the view, so the cold header tank (an unused lavatory cistern, actually) is situated outside the house, up the bank in a stand of thorn bushes. 'What if it freezes up?' I hear you ask. Well, we shall curse a bit until it thaws, but as I have mentioned before, we are near the sea and have much fewer days and nights below freezing temperature than places only a few miles further inland in Carmarthenshire. So far, touch wood, it has not frozen. The hot water tank is a brandy barrel. We call it a whisky barrel, but the pint or so of surplus spirit in the bottom, before we stuck it upright on its platform, was definitely

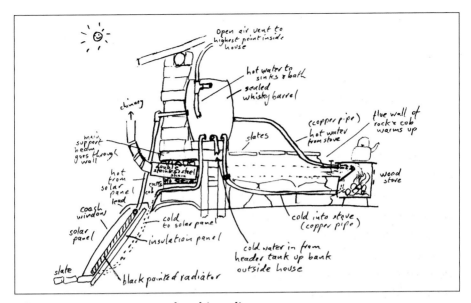

(diagram 15) Hot water plumbing diagram.

brandy. It is made of oak of just over 1" - maybe 32mm thickness – and holds over 60 gals/270 litres of water. This is more than twice the volume of a normal hot tank. I found that a 1"/25mm brass screw tank fitting will fit perfectly tight into a carefully drilled 1"/25mm hole, with a bit of Fernox joint sealer to make it watertight. The advantage of using this barrel is mainly aesthetic, because it means that I could have the hot water store very near to the heat source of solar panel outside the S wall without using up any S wall with airing cupboards and stuff. Thinking long and hard about this interface of heat sources and water store has, I think, paid off. It meant that we needed to have the wood stove also near the S, so I spent a lot of time toying with other functions of this theme and trying to build them in. When the sun heats water in the solar panel, it comes inside the house into the barrel. We tend to assume that we should then insulate it madly, to keep it hot, but I reckon that in our strange climate when so much of the time there might be periods of hot sun during the day, but long cold nights or equally long periods of cold day, it makes just as much sense to use a hot water tank as a low level storage heater. Similarly, it has always seemed a shame to me to see the hot flue of a stove going immediately out of a house to waste its heat outside. Our hot water/woodstove/solar panel system therefore is basically this: *diagram 15* above.

The wood of the hot tank is good insulation, but doesn't keep the heat in - some goes to the house (basically one big space). The solar panel is located outside and just below the hot tank, to give an efficient thermosyphon effect. The woodstove is located about 2.5m/8ft from the wall, facing into the central sitting area. The flue goes from the back of the woodstove up at a slight angle through a low partition wall of stone and mud which warms up from the constant heat of the stove. The wall has going over it the hot water pipe from the back boiler to the barrel, plus three drying lines for airing clothes. It has proved useful in dozens of ways: perfect cat snoozing place; warm bed for seed germination; the drying of logs, twigs and tinder; clothes airing and boots drying. It is located in such a way that the sun in the colder months hits the wall through the window to add some solar warmth. Here's a photo of the wall in use, although, as you can see, not much of the wall itself is visible.

A cat enjoys the sunshine on the hot flue wall.

Grey water

It's no good plumbing in a water supply if there's nowhere for waste water to go, so we dug the grey water system first. I read a bit about grey water and reed beds several years ago and visited the reed beds at CAT before we converted to reed beds at Erw Deg. That system worked fine. We practised not throwing any non-biodegradable substances down the plug hole, and found that we had to rearrange some uses - no cleaning paint brushes with turps. substitute in the sink, for example. Once you get into an organic habit, it's easy to keep it up, so we just continued it here. With that caveat, a reed bed system for a small household is very easy to establish and maintain, especially if you have some slope to play with. The water from our bath and both sinks goes out through the wall (one of those vent pipes at ground level), down under the back path and into a pit about 2m long, 1m wide and 18"/45cm deep. This pit is lined with about 4"/100mm of puddled clay and then another similar layer of topsoil, into which we planted yellow flag irises. A small overflow trench from this pit leads 1m down the bank to a second, identical pit planted with bulrushes. Both bulrushes and yellow flag have evolved to handle having their roots in water for most of the time, so are capable of taking nutrients directly out of water. They clean grey water, in effect.

Preparing the pits for the greywater treatment system.

The planted up reed beds.

One metre or so further down the bank we dug a 10m long swale, of width and depth of 1ft/30cm, along the contour, using the aforementioned bunyip. This swale is intended as a kind of long stop for any liquid overflowing from the two reed beds. We planted it at 12"/30cm intervals with willow sets of golden osier, common osier, and a few super willow cuttings I once was kindly given on a visit to Robert Hart's permaculture plot in Shropshire. The banks of topsoil between the pits I planted with cuttings of blackcurrant, because I have noticed blackcurrant bushes do very well in nutrient-rich damp-earth situations. So that's it. All are growing happily; it doesn't smell, and it looks nice. We don't use enough water to give the osiers much exercise, so occasionally I leave the cold tap on for half an hour to fill the beds up. What a luxury to have so much abundant water. In the photo above you will see three yucca plants that I also planted near the reed beds because nobody wanted them and I felt sorry for them. Still... in a crisis, they are edible.

The only maintenance this system needs is 10 minutes weeding around the edges twice a year, and 10 minutes annually keeping the pipe outlet and linking trench unclogged. One unexpected bonus is that willow tits and other tits like the downy insides of bulrush heads in the spring - maybe for their nests - so we get a good grandstand view of these pretty creatures sometimes.

Plumbing

All the internal water pipes, except 2 metres of copper into and out of the back boiler, are 22mm white plastic Speedfit. This stuff is as easy to fit as Lego. There are two sinks; one for washing up and the other, deeper one for laundry, cleaning veggies and wine making. Experience taught me to go for two sinks with plenty of working space between and around them, and we're happy with it. There. That's enough about plumbing.

Multiple sinks in the kitchen make life much easier.

CAT do some very good leaflets on DIY solar water panels. At Erw Deg I built one religiously according to their instructions: wooden box, 2"/50mm of hay insulation, turkey foil, radiator painted black, glass front. It was brilliant. This one uses the same principles, but is a little less rectangular and is easier to get inside for adjustments. As you can see on the photo of our power station opposite, it is still a radiator, painted black. The size was chosen to fit nicely under one of the big coach windows. Behind the radiator is a piece of polycarbon double channel plastic, with turkey foil stuck onto the top side. This insulates the panel from the ground. I took the radiator to a blacksmith and he drilled out one bottom hole and a top hole diagonally opposite and welded into these holes a hand span length of 22mm steel pipe. The Speedfit pipe fits directly to these pipes. If you want to do this properly, to avoid the oxygen in fresh water rusting through the radiator, and to prevent the panel freezing in winter, you then plumb the piping into a coil inside your hot tank - the top corner going as directly as possible to the top of the coil, and the bottom corner to the bottom of the coil. You also need a vent to let off steam if it boils due to some failure somewhere. This is called an indirect system, and is what all sensible people recommend.

Ours is a direct system, mainly because I know I would have gone completely mad trying to fit a long copper coil into the big oak barrel; in one hole and out of another. So in our system the hot water goes, without antifreeze, through an on/off tap into the hot tank, about half way up, and the cold comes down from the bottom of the hot tank to the bottom corner of the panel, again through an on/off tap. (As the stove boiler is also plumbed direct, for the safety of this system there is an open vent pipe coming from the top of the hot tank and opening at the highest point inside the house). No moving parts; it just works by thermosyphon - when the water in the top of the panel gets hotter than the water in the lower part of the hot tank it just starts taking its place. To get a simple system like this to work without a pump you need the panel to be below the hot tank, so this was a feature of planning the hot water system, of course. The direct system is quicker and more efficient than an indirect system, but we do get a bit of rust colour in the bath water sometimes. From November to March I turn the taps off and drain the panel.

The power station.

This takes about three minutes. To finish the panel construction, I laid a long piece of lead strip, as used for roofing, along the top of the glass and down the two sides, rather than construct a wooden box. The whole panel rests on stones at about a 45 degree angle facing exactly S in a sort of dip we dug for it, which is lined with slates to offer a bright grass-free space in front of the panel. The good thing about this system is its simplicity and cheapness - £10 for the radiator, £3 for the welding, another £10 or so for the bits and pieces, plus £5 for the glass window. (The lead was a gift.) Despite the predictions of plumbers that the radiator on our last one would only last about nine months, it in fact lasted eight years before frost split it at its weakest point. I reckon that this one, especially with the taps and disconnection option which the last one lacked, will probably last just as long. When this one splits one day, it will take maybe three hours' work to replace it. On balance, therefore, I think this is best value (as they keep saying in local government). It works really well, giving hot water on sunny days from April to September.

The hot water system viewed from the sun bit.

Outside touches

*T*he Chinese reputedly have a saying 'man finishes building house, dies' or words to that effect. Similarly, I don't ever see us claiming that this house is finished. There are several features on the outside of a semi-permanent nature that support its continuing evolution as a natural human habitat (he said, reading from a prepared statement). I'll go through them quickly.

Earth sheltering

We built this house into a bank in order to benefit from slope in various ways, which I hope have been clear in this book. All the construction work, therefore, took place in the lee of a looming earth cutting to the NW,N, and NW about 4ft/1.3m from the house back wall. I dug a small ditch round the back with its highest point at the northernmost point, so that it drained from both sides, and watched the bank the whole winter and spring for signs of underground springs or waterways that would indicate a need for special precautions before infilling the piles of subsoil back against the walls. Although we had plenty of torrential rain, the banks, being mainly of clay, held their shape and happily did not reveal new waterways of any kind. Being totally broke by now I decided to take a chance and not invest in any high tech waterproofing solutions, but to rely on good old damp-proof membrane. The infill therefore looks like this (*diagram 16 overleaf*). The membrane is in one long piece, and covers the entire back wall from just under the windows down to the bottom of the drainage ditch, where it

Infilling the back ditch.

79

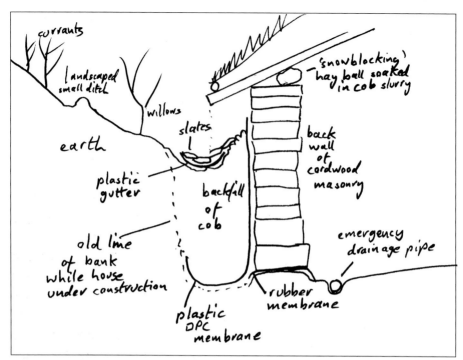

(diagram 16) Drainage details at rear of building.

curves up again for about a foot/30cm to continue the role of drainage ditch underground. We then filled in the gap manually. This took about 6 person-days of heavy labour. Then at ground level I made a first stop drip-catching gutter or drainage ditch, lined with stones, with its highest point at the northernmost point behind the house, so that that catches and drains away the first flush of rainwater. I sculpted the slope a bit above the house to deflect flash floods that might swamp the system in the case of a monsoon-like downpour. The last defence, in case all these water defences fail, is a perforated 4"/100mm land drainage pipe that runs round the entire back wall just under floor level about a foot inside the house. If this ever proves to be necessary it will mean that I have to do more to stop water coming in behind the membrane or under it - probably by installing more land drains half way up the back wall and below the old ditch level - but at least we wouldn't be paddling through puddles to carry the work out. So far the walls appear to have dried out, although the bottom 3"/75mm or so has a few interesting looking mushrooms down in the dark cellar bit where I keep the birch sap wine.

Rubber shingles

Tractor inner tubes are one of the great unexplored rural resources. So far the best use I have found for them is as rubber shingles on the sides of wooden structures. To get them, just ask at local garages and tyre places like ATS. You have to promise that you won't burn them, then you stack them on the back of your milk float, take them home and cut them up. I first cut right across the tube with two cuts about 1ft/30cm apart, making a ring. Then cut at the top and bottom of the ring to produce shingles about a square foot in area, but not quite square, because of the curve of the tyre. You get from 10 to 20 per tyre, depending on the tyre's size. These shingles are very easy to nail onto a wooden surface, lumpy or curved though it be. Start at the bottom with one layer, overlapping each shingle, and using galvanised nails. Work up from there. They are completely waterproof so I use them on exposed surfaces. The only place we have them on this house is on the bit of west facing wall above the bank, covering an area of about 3-4 sq metres. I suppose in time they would harden and crack under the influence of UV light, but the shingles I used on the Cone 6 years ago are still doing fine. I think they look great.

Rubber shingles and edible planting.

Planting

Although to me the ability to completely cover and conceal a building with plants - preferably useful ones - is one of the most satifying and exciting aspects of building my own den, I accept that it may not be a priority for you. Even if it is, your taste is probably different from mine, so I will leave it for others to detail all the types of plant that could be used. Suffice to say that there is a small but growing movement in favour of 'edible buildings' that I am very excited by, as well as a strong body of knowledge being built up by practitioners of permaculture, and that this roundhouse, in its way is part of that movement. A simple plan of the plantings so far is given in *diagram 17* but there will be plenty more plants to come.

Essentially, Jane and I are not bothering with growing any annuals here except the odd bed of potatoes, as the community gardens up the top are so fecund. What we have here are the beginnings of a forest garden of perennial fruiting trees and shrubs, that we

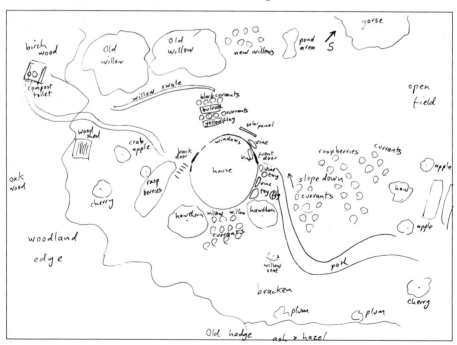

(diagram 17) Planting plan for the immediate area of the roundhouse.

give very little time to but that will in the future, we hope, yield abundantly. I am such a speedy person that the exercise of patience, which is basically what a forest garden needs, translates to me as forgetting about it altogether for long periods of time, so that is what I'm working on doing. There are a few exceptions to this general rule however, as at the moment I have two favourite plants for climbing and fruit production: grape vines and tayberries. There are two Brandt vines outside and one Black Hamburg, and another Black Hamburg growing inside against the S double window that I have trained to go outside through a small hole in the cob and wend its way along the eaves. I am also growing some strawberry flavoured grapes from cuttings from a friend's vine, that will shortly be adding to the outer and inner selection. Ultimately I see vines trailing all around the eaves from the SE to the W, bearing great tresses of delicious fruits. This is just possible, given the trend in hot summers that we have been experiencing. If this roundhouse were not in such an environmentally sensitive location, I would also extend the grape growing area south into a conservatory, as we did at Erw Deg, and thence onto a gazebo or horizontal trellis over a patio and outside eating area. My daughter is marrying into a Greek family with just such a kind of set-up half the way up a mountain overlooking the sea in Corfu. Yes, I think the climate has something to do with the size of the bunches of grapes, though.

Back in our green land, let me recommend the tayberry, if you have not already encountered it, as a prolific rambler, bearing delicious fruits, that thrives anywhere that a bramble would. It was bred in Scotland - hence the name - and is about two-thirds bramble and one third raspberry. The fruits are sometimes as long as your thumb, delicious raw or cooked, and make fabulous wine and jams. The plant fruits, like loganberries or brambles, on the previous year's leaders, so it is a good idea to prune the leaders down to 6-8 in number. If you feed the plant with good compost (ours are fed on horse manure and humanure), it will throw forth its new leading shoots after fruiting with incredible vigour. Last August I trained the first of these leaders up onto the roundhouse

Planting extends around the house and even onto the roof.

roof, where it and three others grew at the rate of a foot/30cm per week for ten weeks. The leaders will also take root themselves at their tips, like brambles, so I am interested to see what this roof will look like in five years' time! I will give it some liquid comfrey compost in the Spring and at regular intervals through the summer. Tayberries have raspberries' habit of throwing up new shoots from near the base, too, and these are the ones I look out for each Spring to start a new plant up or give away to friends. If you have a problem with brambles where you live, try a programme of replacing them with tayberries. They're just as thorny and vigorous, but the fruit just leaves you with a big smile, so changes the whole equation. The only other plant I will mention is the strawberries in the turf. I planted a few strawberries in reachable places in the turf last year and they did well, maybe because of the paucity of slugs up there and the higher levels of sunlight. This year I'll add a breed of Alpine that doesn't need any maintenance or thinning out.

Electrics

Electricity is at its most convenient, and probably at its cheapest, when it comes out of a plug connected to the mains. It's also at its most disempowering, because you take it or leave it. Who knows where it was generated? How much greenhouse gas was released to produce it? Was some of it produced in nuclear stations; if so, where will the wastes go for thousands of years?

The electricity produced by the solar panels outside this house is pollution free, and needs no pylons or mains from anywhere else. We have no electricity bills; just the ever changing balancing act of matching in with out, which sometimes is interesting and the spice of life. Other times it's a hassle - yesterday, for example, when I sat down to write this chapter. To explain a bit, this is written on a Canon word processor that I bought after looking hard at the panel on the back which says how many watts it consumes (24) and what voltage it eats (13.5V DC). So I could probably get away with connecting it directly up to a 12V battery, although I haven't dared. Instead I put its power pack into a 12V/240V inverter that was made locally by an electronic wizard for £25, and connect the inverter up to a 12V leisure battery that I charge up independently of our basic 12V system, so that I can be typing away without the danger that when I or Jane put on a light the whole thing crashes for lack of enough current, and I lose everything I have just typed. This has happened to me twice with typing, which is bad enough, but also used to happen with my digital recording station, where, also twice, I have recorded all six parts of a five minute piece of music - 100 mb's worth - put the light on to make a cup of tea and lost the entire afternoon's work. So this system is meant to handle that, and usually does. Except that yesterday this machine wouldn't recall any file without turning itself off. I got out the voltmeter (an essential piece of equipment in this lifestyle), and found that when this and the solar panel were both connected it read 8.5 volts. Not nearly enough for lift-off. Why so low? I don't know. It's a mystery.* Anyway, I transferred the leads to the main battery, which is two ex-BT 6V gel batteries of 110 amp-hours apiece, and

*p.s. The battery needed topping up with water. I was forced to use intelligence in the end.

which were showing a healthy 12.95 volts. Today, I thought I'd try the leisure battery again - no problem. Everything is back to normal. We have had several hours of direct sun this morning, which will have injected maybe 40 amp-hours into the battery, so perhaps that's the problem solved. Maybe the leisure battery is on its way out. Hard to tell; batteries are funny things. They are almost organic in their behaviour sometimes. They get stuck in certain patterns and have to be jolted out of them. They get old and lazy. Some people love this kind of thing, in the way that a sailor will be looking at the rigging, the swell of the sea, and feeling the strength of the wind at all times. It drives other people mad. They decide that they want to be free of the mains, off the grid, and regret it for every minute until the electricity man is finally summoned back to switch them on again. Being off the grid usually means that you can't have a 'normal' fridge, or a washing machine or a dishwasher. A normal PC is greedy, too, so you have to think about laptops, etc. Or buy a solar roof, which currently costs about four times the cost of this house. While we wait for photovoltaics to be espoused as government policy, so that prices drop dramatically, simple systems like ours are possible, but it is worth considering which kind of person you are. Do you want a technical challenge or do you want it easy? Find out before you jump.

Jane and I decided to be free of the mains here because we had no choice, but we had experience of living with a hybrid system (solar panels and wind generator making elecricity for lights, but retaining mains power for the workshop). The system in this house uses only solar panels, so is quiet and relatively simple, but does make us dependent on the sun. As we have the use of three panels, which are capable of generating about 20 amps, in theory, we have all the electricity we need for the months April to October, and have to be mean with it in the winter. The panels feed the electricity into the battery via a BP Solar controller which is a small box with lights and a tiny brain. It is designed to serve humans with electricity, but in its little robot head its main thought is 'I must Protect my Batteries'. In the long term, of course, this is a useful thought for it to have. In the short term it means that on the third cold wet evening in a row we turn on a light and without warning are instantly, silently plunged into total blackness. The smug little controller is sitting there with a new red light winking at us, and we can't have any more electricity

Two solar voltaic panels on the South side of the building.

until it has had more sunshine, so there. During December, when it is dark for about half of the waking day, we use candles a lot. On balance, this is fine. We can obviate the need for candles by charging portable batteries up at the community workshop on the main system, which also uses a water turbine and a fine Zimbabwe-made wind generator, and wheelbarrowing them back down the track. We don't have a small wind charger because it would be visually obtrusive, and also because unless you have a site that is subject to high and steady winds for most of the time, small wind turbines are, in my opinion, a pain in the neck. Too many things can go wrong. (Murphy's Law.) The wiring in this house is similar to that in a normal house. Brent wired it up for us using 2.5mm cable throughout, which is normally reserved for power. This is because electricity loses voltage in proportion to the length of wire it is required to travel down and the thickness of the wire. A few volts off 230V won't affect lights or appliances much, but a few volts off 12V and you don't have many volts left. So you need to use as thick cable as possible. We have five lights in our simple lighting ring. The bulbs are made by Active Circuits Ltd, and use the same fittings as 230V bulbs, but are of considerably less wattage. The main room and entrance have 20w bulbs, two others are 10w, and our reading light over the bed is 5w. The bulbs last a long time, but most people would view the light levels that we use as decidedly poky. There are three power points around the place too. These

The electric car project.

also take 12V only, and have the earth pin blocked off by a plastic wallplug. Any plugs fitting the sockets have the earth pin removed. It is the system used all through this community, so that we cannot mistakenly plug a 12v item - say a table lamp or a radio - into a 230V socket, nor can we plug a 230V hairdryer into a 12 socket. It is easy to convert many electronic goods and gadgets to 12V, as often they are battery compatible and run on low DC currents anyway. For other gear we use inverters or do without. In this house we have two small inverters for use with the two batteries, to drive the recording station, the CD player and this processor. We are lucky, also, in this community to have Brent as one of our members. He is one of those people who loves the creative challenge of keeping quite complex systems of renewable energy in balance, so as a community we are free from the mains. But it takes a constant vigilance to maintain solar, water and wind systems in balance through all seasons of the year, especially as we are always entertaining new ideas for development, such as the electric car we have built, which runs on renewable energy. In a sustainable future, our society must take account of this need for a small team of flexible engineers to service renewable systems for each village or group of settlements, if we are not willing to depend on centralised sources of supply.

Wood

What wood you have at your disposal depends, of course, on where you live. We are lucky enough to live on a farm that has a third of its area still in deciduous woodland, so we, as a group, can attempt to use these woods sustainably to keep our fuel bills to nil and to be CO_2 neutral. You are CO_2 neutral if the equivalent of CO_2 you put out by burning fuel is taken up by the plants growing on your land. It involves striving for efficiency in burning, good insulation, and seasoning the wood well, so we have worked out a coppicing and wood storage system that we hope is sustainable. 'Coppicing' is the cutting of appropriate hardwood trees in the winter, and can benefit the trees by lengthening their lives, sometimes by hundreds of years. It can also benefit some wildlife such as dormice and some butterflies that have grown used to a coppiced landscape. Good trees for coppicing are ash, oak, alder and willow, all of which grow here in abundance. Birch is not so good - sometimes the stump sends out new shoots; sometimes it doesn't. Beech does not coppice well, but there are only two beech trees on this farm, so they are staying as

Paul with 'Blue' extracting timber at the coppice.

they are. Our procedure is to coppice about an acre per year in a fifteen year rotation. We work in a group on Monday mornings through the winter, cutting the trees with axes and double handled saws (no chainsaws - so much more peaceful). The big trunks are hauled out of the coppice by horse and we hire a planking machine and operator to plank them up for various uses. The smaller trunks are put by Emma and Paul onto the cart and carried up to the communal woodshed. About three cartloads a year are delivered across the field to Jane and me as our supply. We put it in our own woodshed, whose main claim in my affections is that it is a simple example of 'biotecture' - a growing building. To build it, I selected four coppiceable trees about ten metres from our house and pollarded them (cut them off at shoulder height) or laid them so that the upper part of the trunk would rest on the pollarded top of one of the others. Onto this live framework I laid a series of coppiced lengths of ash and hazel about as thick as your arm. Onto this is

Logs awaiting collection.

stapled a layer of strips of pig wire - that kind of fencing made up of strong galvanised metal, the holes forming rectangles and the fence strips being about a metre wide. We have got lots of second hand lengths of this stuff that was in field boundaries and was taken out before proper hedge laying. Its purpose on these roofs is to spread the weight of the turf above it. On to the pig wire I placed a layer of DPC plastic, left over from the backfill on the house, and a layer of turf. I have made about eight structures with roofs like this, and I can recommend it as a very cheap (bag of staples: £2? that's it!) but good looking shelter. The effect is a bit bumpy, but the grass grows fine. If it gets a bit thin, chuck some more earth on top or a big pile of comfrey leaves and it will have a new lease of life. Our woodshed holds maybe one to two tons of seasoning logs for a year, and is growing at about an inch a year! Good thing about growing tree supports

Corner detail on the living woodshed.

is that you don't need to worry about them rotting. Every time Jane or I feel like a bit of arm exercise we cut up some of the logs in the woodshed, split them and stack them against the side

The living growing woodshed, made by pollarding four trees.

of the house, where they will stay for another year or so before being brought inside; first to the woodstore by the back door and then to the drying pile right against the flue wall by the fire. The other categories of firewood are smaller stuff. Hardwood shavings from my woodturning activities are fabulous for reviving a dozy fire or for kindling it afresh. We keep one sackfull by the front door, one against the flue wall and the third, in prime tinder dry condition, right by the fire against the flue wall. Similarly, we keep four boxes or sacks of twigs, preferably ash twigs from the coppicing operations of two years previously, in a drying sequence near the fire. This may seem to you an excessive amount of space and labour devoted to maintaining a good wood supply, but I assure you that it is only the minimum we have found to be necessary to avoid hours of blowing a reluctant fire of damp wood into life. There are lots of folksy rhymes about which woods burn well, and which woods you can burn green, but even if you can get an ash log to keep burning when it's green, it is still using most of its calories to boil off the moisture in its cells, and it will leave your chimney flue running with black gunk. Far better to have a system of wood seasoning so that you can have dry wood in abundance. Even elm will burn, after all, if it has been stored in the dry for ten years. One of the sobering effects on me in maintaining a reasonably workable system of wood cutting and seasoning has been to highlight just how much we take for granted if we just go on burning fossil fuels in an unsustainable way to heat our homes. How on earth are we all, fifty million of us on these islands alone, going to reduce our dependence on fossil fuels without some radical restructuring? I suppose that rural areas could devote a third or more of their land area to forest gardens and hardwood coppice, with groups of people coppicing every week in the winter, and combined heat and power schemes burning the biomass. That's just to get to be CO_2 neutral. Their homes will need to be extremely well insulated, and make full use of solar passive design. The mind boggles attempting to envisage cities also using schemes like this. Cities will need even more radical design. Will they do it in time?

Compost Toilet

When you flush your loo you mix human wastes and toilet paper, with water that has painstakingly been purified up to drinking quality, to make a pungent mess called 'black water'. Where does it go then? Well, if you live in London, until a few years ago, it used to go into a barge and then be dumped into the North Sea. 28,000 tons of it every day. Maybe it still does. Maybe by now European standards have forced the sewage people to filter it, clean it, treat it with ultraviolet and recycle it. In any event, it is an extremely wasteful and expensive way of disposing of something that would be of great benefit in the garden, if composted safely. If you live in a rural situation, therefore, it is much better to be part of the solution, rather than part of the problem, and recycle your wastes directly.

There have been several excellent books on compost toilets written in the last few years, so I refer you to *The Green Shopping Catalogue (see Appendix)* for more detailed choices. I have built a few; they have all varied in funkiness and finish, but all do the same job. The one up by the main house at Brithdir Mawr cost £700 in sawn wood, concrete base etc and is designed to handle deposits from all our visitors to the hostel,

plus regular use from residents working in or near the yard. It has two large (2 cu. m) chambers, a system for catching the pee separately and diverting it onto straw bales for use as rich compost in the gardens; there are large air vents to the chambers, and a decorated toilet room with fancy door, mirror and a picture on the wall. Down here, *semper ut funquet*, we have the simplest compost loo I have ever made, costing £5.

Brithdir Mawr's compost toilet.

Broadly, there are two kinds of composting toilet. Anaerobic gives you decomposition without air but usually with water. A system like this usually needs to be waterproof and tightly joined, and produces liquid fertiliser and methane gas, which can be captured and contained for use. An anaerobic system works better with fairly large quantities and sometimes requires a small amount of heating to maintain the biological breakdown process. Some farms are seeing the benefits of treating their slurry in this way, and there is probably a great future in anaerobic treatment of human wastes in towns and cities. The other, simpler, system is aerobic, in which you encourage the air to get to the faeces and you try to keep too much liquid out. A different kind of bacteria work on the turds, as in a normal compost heap, and worms can also help the process. The heap heats up a bit, and the end result is a dry, crumbly compost that is really beneficial in the garden. This is the system we have here.

Construction

The basic frame of our loo is made of second hand 4"x 4" (100mm x 100mm) softwood that I saved from a builder's bonfire a few years ago. The loo sits on a slope of about 1 in 4, and faces up the slope, so that the deposits can roll down to the back. I used the softwood to make up a frame of two compartments about 75cm square and 1m average depth. These compartments are lined with pig wire stapled to the wood to form a kind of cage, open at the bottom and top. Each compartment is finished off with DPC plastic tacked to the frame. The plastic fits snugly on the top of the wooden frame, and curls round at the base, still leaving an open area of soil at the base of each compartment. This is the best way if you are not going to have too much use. Soil has quite a capacity to absorb and process liquids - the rainfall around here is about 1.6 m per year - so without the shelter being there the soil in each compartment would have absorbed about 900 litres of rainwater. Suppose Jane and I deposit a litre per day of pee each in it. That's 730 litres, so you would expect the soil to absorb that amount of liquid with no trouble, as indeed it does. It is worth doing a few simple sums like this before you decide on the size of your compartments though, just to check that your figures are in the same ball park. If there were four of us living here, I would

Our compost toilet.

have made the floor area of the compartments twice the size. The other consideration in design is that the compartments are big enough not to fill up before a year has passed, because a full compartment needs a year to stand and compost before you empty it. My rule of thumb for this, based on the experience of several years, is to allow about 0.6 litres of settled solid per person per day. (To work out litres, use metric throughout, and measure in centimetres. So a space 75cm cubed is 421875cc. Divide this by 1000 to get litres, i.e 421 litres. Divide by, say 350 days = 1.2 litres per day. Divide by 2 people = 0.6litres per day)

The rest of the construction is a simple shelter made from oak coppiced uprights with a 'Y' top, supporting a framework of oak, ash and hazel crosspieces. The framework overhangs the loo edge by about a metre to keep rain off our knees. Then the ubiquitous pig wire, plastic, a bit of old tarp for better camouflage, and turf. Actually, the word 'turf' is rather too sophisticated. What I did in fact was to go round the garden with a wheelbarrow and dig up maybe the 30 largest clumps of cooch grass, including all their roots and soil ball, and heft them up onto the loo roof, where they still prosper. The compartment which is full is sealed off loosely with old slats of wood across the top, on which rest the sawdust and wood ash containers. The compartment in use has a real loo seat screwed to two pieces of supporting wood, and any gaps to the sides or the back are filled with oak slabs just placed there. There is no seat cover, to allow a maximum of air into the chamber, and no door on the loo. I told you it was funky.

Use

People say 'how can you go to the loo through the rain in the middle of the night?!' etc, but it's fine, really. It is about 20 metres from the back door to the loo. If it is pitch dark, we take a torch. If it is freezing, we wear a poncho. If it is pouring with rain, we take an umbrella (Geri Halliwell's actually - she left it here when she did that TV programme. I suppose I should send it back to her but...) If it is a bright new morning, which it often is, the little walk in the fresh air is a great contact with the outdoors. Anyway, how many other animals defecate in their own dens?

After each deposit, we throw a handfull of sawdust from the wood workshop (hardwood shavings, but softwood sawdust is OK), and sometimes we also put down a handful of wood ash from the stove. This serves the purpose of adding alkalinity for ph balance, of adding potash to improve the final product's value as a fertilizer, and of stifling any insect activity. We have also been known to empty several sacks of seaweed from the beach after a storm into a compost loo, which broke down fine, and produced raspberry bushes over 2m/7ft high. Be creative - it is a fantastic compost we are creating here. It doesn't smell.

Every week I take a metre long stick that I keep by the side of the loo only for this, and push the pile to one side or the other, forward or backwards, to flatten the material below, then shake in several handfuls of wood ash. This is the only maintenance it needs. At no time does one transport the faeces, until over a year has passed, when the compartment in use is full. Then we remove the planking of the old compartment, dig out the contents with a small shovel, and spread them under the fruit bushes. Tayberries love it. I move the seat over to the empty compartment, seal off the full one, add about a third of a sack of sawdust or straw for a soak, and the cycle begins anew.

A last point. After two or three years of operating this kind of system you start to notice the odd unexpected fruit seedling springing up in your forest garden. Maybe strawberries near the currants or currants near the tayberries. One day I put two and two together. Some of these were from seeds that had passed right through our bodies, through the compost system, and there they were, in business, growing. The cycle is complete. Only eat what you like and it will multiply. How's that?

Emptying the compost toilet.

97

Inside the Roundhouse

This chapter will be brief, even though there are plenty of things inside here. I am basically pleased with this layout of a central area about 20ft/.5m across. It has similar shape to a yurt, with its skylight high above and the wooden roof supports fanning out like an iris from it. The essential thing about the inside is that it is almost all wood, and wood is a very adaptive material. I wouldn't say any of this is finished, therefore, merely that it is evolving and adapting steadily.

Partitions are of wood slabs from a sawmill. Where I sit now I face a series of slabs nailed between two internal uprights up to about chest height. Above that are two shelves made of wavy-edge planks and attached to the uprights by rope and staples. Shelves are amazingly easy to put up, so I have used this system all over the place (*diagram 18*).

To my right is a curtain that stretches from the floor up to within 1m of the roof. It is attached to wooden rings that I turned in the workshop. These run on a dowel of hazel that is fixed into two turned holders as seen in B&Q or similar DIY place. The curtain opens for access to our clothes area. My clothes are on hangers on two rails which are suspended

The main living area with sleeping platform beyond.

from the rafters each end by baler twine. Most of Jane's clothes are piled in a huge old curtain that she has also suspended from the rafters near the back wall, where they are about 1m from the floor.

Next to the desk is a glass partition made by suspending the big glass door that is the companion of the front door in between two uprights. In front of this stands our treadle sewing machine. The next space is occupied by the spare giant coach window, which makes a good partition while still letting in plenty of light and allowing views out through it. Jane's desk stands

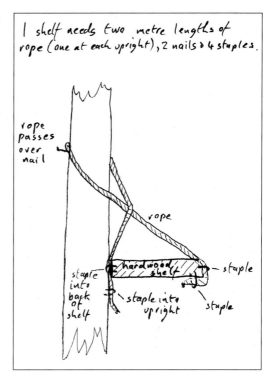

(diagram 18) Shelving detail.

in front of this. The next gap to that, facing SW, is the main doorway into the central area, and that can be closed by a curtain suspended on curtain rings. This time, when I put the rail up, I realised that I didn't need to turn a fancy holder for each end of the rail. I drilled a shallow hole in the sides of each upright and the curtain rail just goes in that. Dead simple. Next to that, in the S, there is a kind of double sided bench seat made by screwing a log of arm thickness to the two uprights about 50cm/1ft9" from the floor, and putting a pallet between the uprights. A futon, covered in a large rug, flips over the bar and rests on the pallet either side of it, thus forming a seat that faces inside and also outside into the sun bit. For the floor, I put planks down in the central area over a framework built up on wooden blocks which were surplus from a local sawmill. The framework needed to be tailored to the floor, which is still the original clay, beaten hard; the floor slopes a bit but the wooden floor does not. Beyond this, in the outer circle, the floor is still beaten clay, covered in places with carpet or coir matting. We take these mats out periodically

The sleeping platform, reached by a short ladder.

on a hot day to air them, as the floor is still cold to the touch. If you are a fussy kind of person, you might call it damp. The outer circle is at a lower level than the raised wood centre, obviously, so it gives the house a slightly split-level feel, which is quite nice. The bath is tucked away down by the flue wall in the lowest part of the SE, so is not visible from the central area unless you crane your neck and look for it. The kitchen storage is made with shelves between uprights. The sleeping platform works perfectly - the view from it is a beautiful panorama.

I am totally happy living in a round house. It is efficient in space, the view is great, there is plenty of light and it just feels completely natural. I have looked at the ceiling and walls periodically and imagined them covered in plaster board and decorated nicely, and yes, it would look like a million dollars. But the natural effect of wood and mud is very comforting, and to get plasterboard over all those angles in three dimensions would be a hugely expensive task. If you ever build one of these, and decorate it all just so, do let me know - I'd love to see it, but we're happy with this just as it is.

Kitchen in the outer ring, with loads of south facing glazing.

The bathroom in the outer ring.   *Instrument storage area.*

Rustic curtain rails between the living area and the front door.

Happiness is a warm rug.

Endpiece

I haven't talked much here about living in a community - it is really the subject for another book. I am convinced that the way forward for us humans is to live in sustainable communities or ecovillages. Whether all the land is held in common in these ecovillages, and what degree of personal property is found to be the most sustainable, largely remains to be seen. By sustainable here, I mean socially sustainable as well. We all have to work within our genetic and social limits, even if we would like to live as perfect communist buddhists. Private space will always be necessary. Here in Brithdir Mawr we are experimenting with setting up a trust which will own all the land and the buildings. I can imagine that for many people a more acceptable set-up might be one in which each homestead were owned privately with a bit of private space around it, and staple gardens, fields and woodlands were held in common. At present we can see that many of society's problems have come from the inefficiencies resulting from each family wanting a detached house, a car, a washing machine etc., and it is true that

Turning wooden bowls in the wood workshop.

Everyone plays their part in communal activities such as haymaking.

intentional communities are far more efficient in the use of resources and in sharing such things as workshops and vehicles. There are other less tangible benefits, too, in living in a community. For me it is the friendship of a dozen others of like mind, the sharing of meals around a huge table and the contact with kids that I just would not have had living still in a nuclear family. But it also changes your feeling of identity, your ambitions, the way in which you view work, money, contributions and free time. How free do I want to be? Well, personally, very. Maybe it will take me years to find my true place in an intentional community, and then my true place beyond that in the wider community of our society.

I do know that without the support of my friends in this community, this house would not be here, and we could not have built it. To that extent a community has been essential for the very existence of this house. It will be interesting to see if the planners finally say yes to this roundhouse, or if at the end of the day it has to go. If so, I trust you will have taken to heart some of my experience in designing and building it, and will keep this old yet new art alive in your own way. Power to your elbow. Thanks to you for reading this, and thanks to Mother Earth and Father Sun.

Ho.

10 Years Later

*T*his Chapter is written in 2007, ten years after we started building the roundhouse. What has worked? What proved unnecessary? What has turned out to be difficult and what else have we added?

The Strawbale Den

My objective with low impact building is to build something that is quick, cheap, of natural materials or common resources, outrageously warm and easy to heat, and is almost invisible from about 50 metres away. Our roundhouse is still standing, of course, but sometimes it is useful to have another space to put up friends, family or wwooffers. I am still experimenting with attempts at the perfect rural bedsit. The one we put up in 2003 is a very low-impact strawbale den that cost £220 all in. It houses one person, or two good friends; it has no running water or mod cons but is well insulated, and has a neat little wood stove, so withstands the coldest weather. It also boasts a magnificent panoramic coach window, and looks out onto a very peaceful woodland glade. This is a brief account of how we made it and some feedback on its design.

We started, in a flat glade in the woodland edge about 50 metres from the roundhouse, with 25 pallets, bought from an agricultural wholesaler at an average of £1 each (some were £2; some were free).

Then straw bales - as tightly baled as you can find. Buy them around summer time when the farmer is pretty sure

The strawbale den in the woods.

that he won't need his whole stash from last year and he needs the space for this year's, and you might get them for less than £1 each. We bought 60 at £1 each. Organic, as it happens. Costs are mounting, but as yet they are handlable, wouldn't you say?

We have helped in the building of several huts using straw bales either as infill in the walls between posts, or bearing the roof load in their own right. This hut has no wooden frame, so I preferred to place the bales flat, not on edge. This means you use up more space and more bales per metre height, but it also means that your walls are good and stable and 18"/half a metre thick, so we are talking serious insulation.

The shape of the den here was dictated largely by the trees nearby, and is egg-shaped from above, not circular. Here is the first course of twelve bales. If you are nervous about rodent visitations, you can put a layer of chicken wire around the bottoms of the bales and about a metre up the sides at this stage.

I cut dozens of stakes of poplar and hazel, and sharpened one end of each, helped by a gang of boys from a camp nearby. Each bale is pinned to its position by two stakes, which are thumped by a big mallet until they go right down into the ground. Actually the best mallets to use are so big that the correct word is a maul - a noun one doesn't get the chance of using very often. The bales are also tied to each other by baler twine.

Next you have to make a few half-bales, so that the second layer overlaps the first, as bricks are laid. To do this, you take a bale and, without cutting its twine, thread new twine through the bale in its middle

The first course of strawbales laid onto the recycled pallet base.

next to one of the existing pieces of twine, with a new length of about 4 metres of twine attached. Professionals use a special tool for this - a metal or wooden bale needle - that is about 8ocms long, sword-shaped, with a notch or hole near the tip unto which you put the twine. If you get the idea, you can make up your own from that bit of the kids' buggy in the garage that you never got round to fitting back on again. Or make one

Wall support near the door opening.

from wood. With a hut this size, it doesn't have a lot of work to do.

Now we have our half-bales, we can proceed with the next course of bales. I decided to make this hovel particularly low, with the intention of having no furniture as such, so the window is set on top of the first course. Several people helped us with these wall and roof stages, so it was quick to do.

As a sill we placed planks of oak, that I had set aside from the last visit of Dai the Woodmiser, on the first course of bales, and pinned the next course of bales on the first until we came to the sill, when two people held the large coach window in place and others set the bales tightly on either side. The window is, luckily, almost exactly three bales' thickness in height. We continued setting the bales, pinning them with stakes and tying them together. As an additional strengthener, I also fix stronger poles every three bales or so, about two metres long and about as thick as your wrist, firmly hammered into the ground on the outside of the structure, tight against the bales. I then hammer a twin pole directly inside, close up against the wall, and down through the gaps in the pallet into the ground. The poles are then linked by a piece of baler twine through the wall. We then pulled the twine as tight as possible and tied the twine to the poles. Above is one near the door. This picture also shows the next stage after all the four levels are tied - the fixing of a wall plate of oak logs all around the top for the rafters to rest on.

We chose slightly curved logs about as thick as your arm, and lashed them tightly together to form a strong circle. We fixed them to the bales with coppiced hazel rods of about finger thickness – the kind that hurdles are made of. If you take a rod about an arms length, you can twist it and bend it at the same time so that the fibres separate a bit but the rod does not snap. You can then twist and bend

Window with double lintel and wall plates. the ends right round together so that you have a kind of wooden staple about 30cm/a foot long. I hammered about 15 of these staples into the top bale from above, one pin on either side of the wall plate, thus fixing it in position. The wall plate, when it reached the window, divided into two beams, one on either side of the glass. The glass is held tightly in position, but does not extend up above the wall plate. The roof rafters could thus rest on it all the way

round, and we had a good double lintel for the window, as in the roundhouse. In the previous picture you can see that we also doubled up the wall-plate over the door, for strength. Above you can see the window firmly in place, sitting on the oak sill, and held at the top by the double lintel. Notice also the wall support poles, as above, on either side.

Next we put up the reciprocal frame of some hardwoods and some pine rafters. Each rafter

Tying the ends of the reciprocal rafters.

was about 5 metres long, and we trimmed them after fixing the frame. Here's Gareth tying them with baler twine. There are 13 rafters – there is still one to fit in when this picture was taken.

Making a den with a low roof is much easier than making a larger structure. All Gareth needed here was a small stepladder to reach the central hole. If we needed to adjust the rafters a bit we just held them up while someone moved the ones underneath. Our roundhouse is over 10 metres/34ft across inside, but this den is about five, with a low ceiling of just over 2m, about 7ft, at the skylight in the centre. Everything is within reach to a tallish person. It's worth considering just how small you can tolerate a den, because everything becomes so much simpler. We had

this structure built in three days, including the roof frame. We could then cover the whole thing with a tarpaulin to protect it from rain, and work at our leisure on the radials, insulation, turfin', wall plastering, etc.

(Feedback note - On the other hand, four foot high walls, ie four bales high - have proved to be a bit of a limitation. It is especially irksome bending double to get through the door when you are carrying an armful of logs. You can stand

Skylight and rafters from underneath.

erect in the hut, fine, and many guests love it with a low ceiling, but if we could replay the past, after four years of use, we would have put an extra fifth level of straw bales and another window, too.)

For radials we used slab wood, because that's what we had. On top of that are canvas and some old blankets, then wool, then straw, then two sheets of silage plastic, then turf. The skylight is yet another coach window. Why fight it? They are strong, optically excellent, curved to let the rain off, and are safety glass. I got four for £5 each, but you won't get any more from the same bus depot. Better try one near you. Here's the roof from underneath.

Jane putting on the first row of turves.

Eaves detail showing galvanised wire straps.

Raw wool used as under floor insulation.

And here is Jane putting on the first row of turves to hold down the plastic. Don't cut the skylight hole in the plastic before most of the turves are on, because the plastic could move or stretch under the weight of the turf, or you could be kicking yourself for having a hole not quite in the right place. The coach window simply rests over the whole central area, is held down with a few turves, and is draught-proofed from underneath with raw wool.

The turves are faced at the eaves, as with almost all my dens, by logs of about an arm thickness, resting on nails hammered into the ends of the rafters. They are fixed together in a ring by short lengths of galvanised wire held to each log by staples. The plastic is folded and pinned at the back of these logs.

Insulation is necessary in various gaps – in particular between the rafters and under the pallet floors. I used raw wool for this:

The stove pipe is of double, insulated stainless steel, that our son-in-law bought at a car boot sale, complete with fancy cowl on the top. If we had bought it new it would have tripled the cost of the whole den. So if you see some second hand, buy it. You can build the den later.

Where it goes through the roof, it is further insulated by an old fireguard fixed around a double layer of fibreglass fire blanket. Where it comes out I made a cone of metal to deflect

Insulated flue pipe going through roof.

the rain, and also did a lot of messing about with rolls of metal flashing tape.

It works in keeping stove-pipe heat from the roof, and water out. The woodstove is lovely and warm, and will just take a pot for an early cuppa.

(Feedback note: We had to renew the flashing round the flue in 2007 when a leak started near the flue. There was a sweet little bank vole's nest under the cowl and a handy stash of hazel nuts. I mostly call this den a hut, but we are considering an upgrade to a hovel, which has a stove in and on which you can cook, with maybe access to running water.)

At any time after getting the roof on, or even before if you are confident about good weather, you can start putting mud on the walls. What we used was a mix of about two sand, one clay, and one horse manure. Horses are not just for looking good and pulling things. (If you're doing this in a town, or have access to one, try human hair from a hairdresser's if you can't get horse manure. I hear that it's very good, and keeps the wild boar away.) Some people just love smearing this mix over straw and could do it almost indefinitely.

Flue with flashing and cowl.

Jane smearing mud mix onto the strawbale walls of the den.

Tony and the cedar bark cladding.

Here's Jane smearing mud mix round the door.

And here it is very nearly finished: After three months or so I reinforced the mud on the south and west facing walls with some cedar bark that looked too good to waste. I pinned it on with pegs about a hand span long made of willow. Now it looks like this: One day it might be a good idea to do a final render of lime putty.

Roundhouse Feedback

I came across an old photo the other day. Jane is sitting in a rickety chair, dressed in layers of mud-covered clothes, eating a bowl of muesli. Around her is chaos. Piles of wet logs, straw, a wheelbarrow of mud, a blanket where a door might be and clear views through non-existent walls of a bleak January windswept mudscape. That was us ten years ago, when we moved in because we could at least light a fire in the milk churn and huddle round it rather than crawl into our damp bender and ask 'how much longer?'

Now, in the spirit of good permaculture enquiry and search for useful feedback, what have we learned that might be of use for the future, or for you if you are on the same mad path?

So I'd like to go through some feedback in two chapter categories - Physical/ design/lifestyle and the Planning Saga.

Physical/design/lifestyle

Visual aspects

Visually it's looking better than ever, with grapes and Japanese wineberries on the roof, willows to the north, the reedbed healthy and trees growing around it. A friend of ours, Mike the Mill, has a light aircraft that he flies over in occasionally, and he reports that from the air the house is increasingly invisible.

Two elder trees grown by the reed bed from cuttings in 2002 now produce plentiful flowers and berries for wine. There are three Black Hamburg grape vines growing along the eaves and spreading onto the western quarter of the roof, producing about 15 kilos of grapes for wine in a good year. There used to be four vines, but one died, being situated too close to the bottom of the flue where it came through the wall. This sometimes leaked some tar, which I assume was unhealthy for the vine. I prune the vines fairly severely in the winter.

Round is good, and turf evolves well. It pays to design solar passive aspects in, and they have worked well. The planners don't like the look

The roundhouse 10 years on just looks more and more natural.

of the funky shiny external chimney flue, and there is a safety consideration in that the top of the flue is dangerous to reach. In January 2007 I fell from a ladder trying to clean the top of the flue of soot deposits, and was out of action for three weeks. With this in mind, we are redesigning the chimney flue, see left.

Radials being fitted on the roof at Eco Casa

Another visual feature, that came up as an issue in a course building a similar design, is the look of the eaves when you place secondary rafters, as here, coming off the main rafters in parallel. Because these rafters cross the henge at a slightly different angle the shorter they get, this gives a wavy edge to the eaves of the house. I don't mind this - I actually like it and it breaks what would otherwise be a level line at the eaves, but if you want a neater look to the edge of your roof you can place your next level of roof supports across the rafters at one hand span intervals, like a giant spider's web. We tend to make our roofs like this these days, because these 'radials' each have to do less work than the secondary rafters, so can be smaller and more manageable. Here are the radials being fitted in 2007 on the roof of the Portuguese eco-casa at Cabeca do Mato, as seen from underneath.

Cobwood

Cobwood is great. Half a meter of eaves and the earth sheltered nature of it means rain almost never hits the walls. Don't worry about exposed log ends or mud washing away. It doesn't happen except in very exposed locations, and we eco-builders choose tucked-away corners, do we not?. The wood doesn't rot, and the cob doesn't crumble. It just stays where it is. I was wrong about silt being useful in the mix, by the way. It's no use at all. For draught proofing or occasional cob repairs in places I use Ianto Evans' (Cob Cottage Wizard) formula, amended a tiny bit: three

parts sand, one part clay, one part chopped straw or horse manure.

Bottles sit well in cobwood - place a clean wine bottle in a jam jar, tape them together and place the result clear end out to face the sun, sometime, in the wall in place of a log. If you want a light mandala, plan it out and make a full size paper or cloth template and tack it up on the uprights so that it hangs behind where you're working. Mark out on

Placing bottles into the cobwood wall.

the template exactly where the blue or white or whatever bottles go, and fill the rest in with cobwood. The more exact you are, the better will look the finished wall, inside and out.

Cobwood is stronger structurally than straw bale and is more resistant to rat attack. In all these phrases hangs a tale. When rats did find us, after 5 years, they moved into the straw bales in the south wall. The massively insulated roof is good but rats and mice love straw, and rubber is not as impervious to ants or plant roots as plastic is. So we are very careful not to leave any food around, we keep all foodstuffs in metal boxes, and we have put silage plastic over the rubber under the turf. Everything gardens. Yes, even in your roof.

People have asked me what types of wood are best for cobwood walls. I haven't

Mandala pattern of bottles.

found any wood that wasn't fine. You don't need great strength - all the wood is doing is holding, by compressive strength, what is above it up to the henge. Probably the best wood would be the most insulating, i.e. the wood with most air in it, so soft woods like pine are fine. Aspen, poplar and cedar would be good. In the Clynfyw roundhouse, whose mandala is shown on the previous page, you can see very tasty cobwood walls. These are very old laurel trees that could not be burnt because of a high cyanide content (!) but are great as cobwood. Most wood will split if you use it too green, so if you want to avoid major splitting, cut your wood at least a year in advance and store it in the open under cover but well ventilated.

A common question I am asked, by people worried about mixing wood and mud, is "But doesn't the wood rot?" to which I reply "Do books in a library rot?" or "Does the wattle in wattle and daub rot?". Wood rots if it keeps getting damp. These walls stay good and dry. If they don't - worry.

Roof

You will remember that our roof is canvas, then straw bales, then rubber 1mm pond liner, then newspaper, then turf. After eight years we had our first drip. Wailing and gnashing of teeth. The water would surely creep all through the straw, soak the entire roof which would then start composting and spontaneously combust. End of story - Roundhouse Drama Ends in Mystery Blaze Shock. End of the World. I have since read five Global Warming books in a row, and recommend this course of action should you wish to retain a sense of perspective on personal potential disaster. I still don't know what causes the holes in the rubber - they look carefully worked at, and are rarely more than 2mm across. It could be dock or couch grass that pierces the rubber, or it could be ants that make the hole first. In any case, there are solutions:

1. the drip occurs directly below the whole - the water goes straight down through the straw. I find the hole, mend it with a puncture repair kit, and the drip stops.

2. Locating the hole: you need a long 3-metre stick and a helper. The thing is, it is easy to locate a hole with all the rafters, windows, cobwebs and other paraphernalia indoors. On the roof, however, it is another matter. Lovely

The Eco Casa roundhouse at Cabeca do Mato in Portugal.

view. Nothing but Earth below us / above us only Sky, to quote the classics. The central skylight and the long stick resting against and up beyond the eaves can, however, be seen from both locations, so we use this as the common markers. I stand inside and yell to my assistant "left a bit; right a bit..." as he or she moves to line up skylight, drip and stick seen as through the windows. I then estimate the distance from the central skylight to the drip down this line. Up onto the roof, the stick still standing in the same place, I estimate the same distance from the skylight on the turf, and start digging. (Not with a spade, no.) Usually if we clear a square meter of turf the hole is clearly visible, and is repaired in ten minutes.

3. When building a house with a turf roof, use more than one waterproof layer, whatever that layer is. Now we have an agricultural plastic, such as used for silage stacks, under the turf over the rubber, and leaks have stopped. It's a shame in a way, because we wanted to use all-natural materials, and plastic is not natural. But it is readily available, and approximately one twentieth the cost of rubber, so, hey, sometimes we have to compromise... Nowadays I use a layer of terram or similar geotextile - the kind they use under roads to resist roots and shoots - plus at least two layers of plastic, plus rubber if possible. The project I was involved with, at the eco centre Cabeca do Mato in central Portugal, has two layers of plastic, then 1mm rubber, then one layer of plastic (two in places) then cardboard, then soil. This, planted up with succulents and mulched with straw, will hopefully outlast the end of the world.

Floor

The floor started out as bare packed earth. We got lots of ethnic brownie points from visitors but here it's cold and damp.

It might be better where you live. Also, I didn't put a lot of care into ensuring the floor was level before moving stuff in. Now we have a raised, planked, level area in the central living space with home made rugs on - very comfy - and in the outer ring I've put down slices 6-10cm thick of round sitka spruce, but it could be any wood. Looks good, and much warmer than bare earth. We haven't decided how to fill between the slices, though. Sawdust/glue or sawdust/linseed oil or sawdust/clay are contenders.

We still have a small area of natural untreated earth floor in the south sector by the large windows, and there are geranium plants, some succulents, and a pineapple sage growing in the floor itself by the window. Intrepid bracken shoots still come up each spring under the kitchen sink, too. You have to admire bracken for its perseverance.

Compost toilet

The compost toilet - two chambers about 1 cu.m. each, plastic lined but open to the earth, fed with hardwood shavings and sawdust - is fabulous. The compost amounts to about 9 or 10 wheelbarrowfulls a year and helps a true cycle take place. I know the bushes love us for it. In the winter after a big storm we visit our local beach once a year or so, carrying lots of sacks and dressed in many warm layers of dirty clothing, in a kind of mediaeval peasant pageant, to fill the sacks with seaweed. This adds minerals to the compost toilet mix and smells distinctively nautical for a week or two as it breaks down.

I did get asked by someone to look at their compost toilet because nothing was decomposing adequately in it after several months of use. It turned out that they were using fresh softwood sawdust as a soak, which was so resinous that it was preserving the turds rather than breaking them down. So try to use a soak that breaks down OK.

Wood stove

A woodstove that heats your water is feasible, makes your house heating carbon neutral, and it is good for your soul and genes if you can see the flames through a glass door. To produce adequate supplies of dry,

A switchable flue arrangement allows heat to be sent to a thermal store.

seasoned firewood is one of our major routines, requiring maybe a day a week, and operating over a two-year coppicing/cutting/drying cycle. If people in suits realised just what being carbon neutral involved they'd give us medals instead of distrust. It is an essential discipline for the times of energy descent coming upon us. How can you avoid escalating gas and oil and electricity prices, energy wars, and the 'long emergency' looming? Don't buy into it. Get a good double handled saw, an axe, a hectare of land and a planner on your side. But I digress.

We have found the design of the long flue going through a section of radial wall, before going outside and up, has pros and cons. The Pros are that the wall warms up and forms a very useful internal radiator and space on which to superdry logs before they go in the stove. The Cons are:

1. The planners don't like the look of a shiny bendy metal flue coming out of the wall, albeit by means of an insulated metal junction, and going up the outside of the house. I don't like it much either, to be honest, although it definitely is funky.

2. It is tricky, not to say downright dangerous, to fully clean the flue.

3. Sometimes it is hard to get the smoke from a newly lit fire to realise that the way up is by means of a three metre horizontal tube. Some mornings have been very smoky mornings for twenty minutes or so.

If you are adamant that you want to instal a horizontal flue just like we did, however, do make sure you have an inspection/cleaning chamber at the bend before it goes vertical.

We have just changed our chimney to go vertically up through the roof after about a metre horizontal run. This provides a turbo mode of the flue, to allow easier fire starting plus the option of leaving open the fire doors for added heat and maybe simple cooking on the open fire. We have retained the wood warming/long flue/kakeloven option, however, by having a three-metre diversion of flue down and through a built up piece of thermal mass. This will kick in by the manual turning of a baffle in the flue which closes off the direct turbo option (when the fire is going well) and opening up the thermal diversion. We gave careful attention to thermal insulation of the flue as it goes through the roof, as in the strawbale den.

Renewable Electricity

The photovoltaic panels have continued to provide a smooth learning curve. They are fine in summer. Silent, easy, and still relatively expensive. We still don't have a TV, fridge, toaster, or washing machine, but in the four winter months of November to February there is still not enough electricity from solar panels alone. There just isn't enough daylight of good quality in the winter to be able to rely on solar panels for anything but one or two very low power LED lights. Using the laptop is out of the question after two or three days of rain or overcast skies.

I tried investing in an Aquair stream engine to mount in the stream nearby which would in theory generate 80 watts or so, but it needs a stream flow of about one metre per second or more to generate any significant amount, and the winter of 2006 just didn't give enough rain to do that. I put in a lot of work and money to come to the conclusion that this project wasn't going to deliver the electricity we needed. So in 2007 we installed a small (200watt) wind generator of Chinese manufacture. Even though it is only 4 metres high it really goes well in a decent wind, and has been very useful during the winter. The internet is the best place to investigate what is on the market. One good site is www.alternativeenergysuppliesuk.com but there are dozens.

I am not trying to influence you to go off-grid, by the way, unless you really want to and have good reasons for doing so, because the amount you need to teach yourself about limiting demand, watching the weather and making plans accordingly, keeping batteries in a good state, and understanding 12 volt systems, inverters, etc, is a serious commitment. Even going non electric altogether can bring problems of justifying non-sustainable candles! On balance I am very grateful to have had this experience of living off grid, and will continue to be happy to do so, but sometimes it's no rose garden, and will probably never 'pay back' your investment until civilisation actually collapses, at which point you can be as smug as it is possible to be for the couple of days before other people take over your house. This doesn't have to be as bad as it sounds; I met a Croat in Findhorn who lived on an island just off Dubrovnik and had installed solar lighting in the early 1990's, at great personal expense and to the amusement of most locals. When war broke out in the early 1990's the whole village power supply was cut off for weeks at a time. His was the only place with power for miles. Most of the villagers just hung out at his place every evening just so they could see each other while they chatted.

Reed bed

The gray water reed bed of reed mace and yellow flag - two square meters by half a metre deep, clay-lined, then another identical one further down the bank -works perfectly. Twenty minutes maintenance to clear around the inlet pipe once a year, and ten minutes to cut the dead stalks back in winter. Piece of cake. The only problem with it has been the planners, who say that we have created a new habitat on what was marshy grassland and on what we say was sandy bracken bank.

Kitchen

Two sinks are great. If you do washing by hand and make jams and wines and stuff don't try to make do with just one sink. They're only a tenner second hand.

The solar water heater is good, too. If I had a bit more money I might invest in a bought job though, because these days you can get such efficient solar water panels from places like B&Q that it's probably worth it, especially on the days in Spring and Autumn when it's warm enough to not light the stove until nightfall but you need some hot water. You get warm water with ours, but modern ones would give you hot. I would still avoid a system using pumps if I could, because it's one more indispensable link that could go wrong, isn't it? Who would have thought that one of the biggest problems for people in Tewkesbury in the centre of England, hit by floods in the summer of 2007, would be that the power going off would disable the pumps supplying water to all those homes? Water, water everywhere, nor any drop to drink. (Or to flush the loo - another good argument for compost toilets, preferably on hillsides).

Windows

We are quite happy with the windows - that is to say that they don't leak, and are great for views. We open and close the window in the north sector by the bed, but don't feel any need to open the other windows that don't open. The back patio door still opens wide, and the big glass front door from the Bank is fine. If we want to ventilate the house fully we open both doors at once. We don't use many of the air pipes in the walls, but a great tit has moved into one of them, so they are handy habitats.

The skylight moved once in a storm, and I had to fiddle about repositioning it. I could put some big stones on it, like they do on exposed roofs in Cornwall and Asturias, but I can't be bothered, quite frankly. I hate carrying stones about.

We did try, in the early days, removing the skylight and lighting a fire on the clay floor in the middle of the house, but the yurt-like shape doesn't lend itself to natural chimney functions as a tipi shape does, so the experiment was smoky and short lived.

Design/lifestyle

Some changes Jane and I would make design-wise in the light of experience and eight years' use:

- 400 metres is maybe too far from the centre if you wish to be fully functioning members of an intentional community that shares meals, meetings, work groups etc. It was OK while we were in full health and had no outside worries, but when the planning saga became intense, the physical distance became too great for me, personally, to want to trudge a quarter of a mile up a wet lane in the dark every night to eat supper. It also took up so much time that I could not feel I was contributing my fair share of communal work. So although we can still be said to be members of the wider community aiming at sustainability, we are no longer members of the housing co-op of Brithdir Mawr.

- A greenhouse/conservatory on the house would vastly improve our productivity by some protection for seedlings from slugs and by lengthening the growing season. We avoided one so as to be inconspicuous, and may now be allowed a porch, with which we must be content.

- A workshop on site would be more efficient and reduce transport. The planners turned down my request for a strawbale workshop because I wanted a translucent plastic roof for it.

- A music studio where I can make lots of repetitive noises (it's sometimes called practising, sometimes drumming, sometimes snoring) will be on my design list for the ideal homestead, too.

- We just make do with one open space divided by a concentric ring from utilities, but an office fits badly in this round space, and, if we had small children around all the time, a different design solution would be necessary. Having belatedly read Men are from Mars - Women are from Venus, I now realise how important a separate space is for either one in a couple to retire from all company for a while. Several round pods, maybe, like in traditional African family compounds, would be worth considering.

The question I still regularly ask myself is 'Is our lifestyle sustainable?' The longer version, which is the umbrella under which I have lived for the last 30 years, is 'Is it possible to live sustainably while still living comfortably in this culture and society?'. We were assessed by a researcher from the Oxford Brookes university as having a 'one-planet' eco-footprint, but I still have my doubts that we are sustainable enough. I have spent so much time fighting to keep the roundhouse that although Jane is still 'one planet' I am probably emitting too much carbon to be truly carbon neutral. The answer I would now give to the second question is: 'Only with the support and encouragement of the national and local authorities. You cannot live in a low impact, sustainable, earthright way despite your local authority! Which brings us onto planning.

Planning

I am not, nor have ever been, a planner. In the mid-seventies I nearly got to know the planning system quite well by having my next assignment, as a Performance Review Officer in an English local authority, be to look at the performance of the Planning Department. I didn't do it. I left my job, sold my overmortgaged house, bought a truck and went off with my young family learning fruit picking and washing up in France and Spain, came back and joined a self-sufficient commune in Wales, and have been 'one of them' ever since, rather than 'one of us'. Before leaving, however, I spent many fruitless hours pondering on the question: "How do you assess the performance of Planning?" Oh yes, I know people in suits will tell you that their response rate to applications is a certain high percentage within so many days and so on, but how can you tell if a Planning system is working? What is its purpose? One thing for sure is that different countries have different planning systems, based on how they view the concept of land ownership, the rights of the individual as compared with the rights of the state, population density, expectations of self-sufficiency, and so on. Even very similar regions such as Brittany and Wales have different Planning expectations, and it would be a most interesting exercise to seek best practice on the international level to find the best parameters for a sustainable society in different geographical situations. I have read that the world's only truly sustainable city was what is now Mexico City when the conquistadores arrived. A city based on canals which provided water to homes, irrigation and for transport. Continual dredging of the canals provided rich compost for guilds of highly productive plants grown by decentralised households at the water's edge. I know of no modern city that could claim to be sustainable. London's eco-footprint, for example, is the size of the whole of Great Britain! This would be laughable if it wasn't so desperate.

We often assume that we know it all by now. This is one area in which, in my humble opinion, we are stumbling around in the dark using unquestioned concepts inherited from the Normans. There is plenty of useful thinking to be done, but I am happy to say that almost none of it will be done by me.

There is a lovely sentence in the documents we receive occasionally from our local planners:

'Sustainable Development has emerged as the overarching objective of the planning system in the last decade.'

Wow! Happy days! Objectives are quantifiable, measurable and attainable, or they are worthless. Overarching means that it has priority over everything else - economic need, profit, nice view, winning, tradition.....
So all new developments will be assessed as to their sustainability, yes? Every new planned building, road, school, factory, airport, housing estate, park, marina, shop and casino will contribute to a zero carbon future where biodiversity increases and all humans have a one-planet ecological footprint. Furthermore, the planning system will have taken account of the fact that by 2030 the effects of global climate change and oil/gas depletion will be so severe that many areas of Britain will be virtually lawless, racked by floods, storms, disease and economic depression - constantly being invaded by hordes of homeless people from all places and walks of life, trying to find somewhere that they could call home, somewhere maybe they can grow some food, somewhere where they will not be moved on. (Britain will be one of the least hard hit - some countries will lose most of their useful land to the rising sea levels).

I'm not sure that planners have quite taken this on board yet. The armed forces have, or at least the negative side of it, with their advance planning. In 2005 they asked the oceanography centre in Southampton, which assesses future climate and ocean situations for input into the ICCP reports, to give them an assessment of the likely situation facing Britain in 30 years' time. The reply was that the environment could be so unstable that it would be unwise to rely on the security of ports or airfields for Britain's defence. In spring 2007 the British government announced, without giving any convincing reasons, that they had ordered two new giant aircraft carriers for the navy - the biggest ships that will ever be built by government - at a cost of £4 billion. So someone is looking at the future, even though our modern day Noahs work for the Ministry of Defence. Realistic planning today should include Arks - hundreds of them.

One reason I left the whole game in 1976 was because I realised that I had not lived enough to make sense of any answers I would receive after asking "What is the purpose of Planning?". Maybe now, after 30 years' experience of trying to live sustainably, I might make more sense of the answer, but am one of the least acceptable people to be asked the question. I am one of the biggest thorns in the side of our local planning authority they have ever had, by building an ecohouse on private land without asking them first (perfectly legal to do that, by the way), then resisting pulling it down until we had good, sensible, logical and earthright arguments to do so. We have never had a good reason to, and 99% of all visitors have said "They can't make you pull this down, can they?" or words to that effect. Take a look at the guestbook on www.thatroundhouse.info. The support we have received has been astonishing, and a testament to how many people hanker after a simpler, more natural home and way of life.

The Planning file on the roundhouse stretches from 1999, when the house was discovered, or at least confirmed in its existence, by a spotter plane looking for caravans, through to November 2007, when I write. Our local planners, the Pembrokeshire Coast National Park, have been quizzed by dozens of media agencies including the *Times*, the *Independent*, *The Observer*, the *Economist*, the *Guardian*, the *Daily Express*, the *Daily Mail*, the *BBC*, *ITV*, *Discovery Channel* and many foreign news, radio and TV companies, which have all written or published articles, film news, features and web logs on the roundhouse. The Park's website guestbook was withdrawn after being clogged with messages in our support.

I don't want to get into a rant about any particular situation. The full history of our case, as it happened, is there for you to plough through, if you want, on my website (www.thatroundhouse.info). Jane has now taken up most of the detail with the planners - I'm too jaundiced to be nice to them any more. The saga was for us in colour, in real time and real life, not just words. It might make a good film. Some scenes were quite dramatic - the opening scene in which our hero, taking a break from chopping logs, goes into the big community house for a glass of water. He is wearing his tough leather hat. Maybe he is played by Matt Damon or Bill Nighy. Maybe not. In the kitchen, at the big table, is an unexpected sight. There is a woman with a sharp face, glasses, and power suit, giving an interview to television news cameras about all the planning transgressions that she has discovered in our farm. A dome, a strawbale house, a roundhouse, a lake... There is a shiny car parked right there; not in the car park, but in the centre of the farmyard, guarded rather sheepishly by an oldish man in a worn grey suit that we come to know as Brian the Enforcer. (A cameo role from Michael Caine?) This is an invasion, a blow to the heart. (Stirring music) Our hero leaves his hat on. He marches outside. Will he wreck the shiny new car with his 16lb log splitter? Oh the joy! No, he will buy a laptop, learn computing, and engage with the rest of the community in a dedicated defence of low impact living that takes him to the very gates of Bonkerstown and Burnout Gulch.

The public inquiry would make for good TV (Maggie Smith is the barrister for the Park - "A hundred bottles of wine a year? Huh! I could get through that quantity myself!..."). She is replaced by a younger model, a tough talking ex-miner from Yorkshire (played by Colin Whateley from Morse). Every time we see him in court corridors we invite him to change sides: "Come on Paul, it's never too late to give up." Emma promises him a mud hut of his very own.

The film has good drama - the public demo in Haverfordwest, the planning appeals and some of the court scenes - but I would like it to also include the real moments too - logs cut and chopped with verve, vegetables grown, slugs washed off salad, raspberries picked, music played, walks through the woods and cups of tea shared with visitors. "How long did it take you to make this, then?" is still question number one. Subtext: "Blimey, even I could make one of these! I wouldn't have a mortgage or pay rent. I would be free!".

The purpose of Planning - to set people free? Hmm. OK - here are a few things I have learned in this process and would like to pass on to you. Don't ask me for any more stuff unless you absolutely have to because I am tired of being an eco warrior; of planning the next move; of thinking strategically; of reducing what is real to what is only concepts; of substituting what Edward Goldsmith calls the 'surrogate world' for the real one:

1. I frowned much too much at first. A frown lasts for a month but a smile lasts for years. Be a happy victor not a sad victim. Don't be afraid to ask a public servant, "Why?".

2. Get clear whether you are willing to risk a battle or want to do it the slow way by asking for permission in advance, etc. At present most low impact livers do it first then ask for permission, because planning policies are actually so far behind their 'overarching sustainability' rhetoric. The statutes support you in this by retaining various clauses that allow you freedom from enforcement - different lengths of time for different situations. This is a kind of low impact concept built into planning law that planners don't shout about. If you build yourself a dwelling and live in it, it is immune from enforcement four years after being substantially completed. But you must prove very conclusively that four years have elapsed, that it is the same building, etc. Don't assume this proof is easy - it's not. You must also take a lot of stick from people who think it is illegal to build without permission. I had quite an argument with the presenter on You and Yours of Radio 4 before she checked it out for herself. On the other hand, do not assume that the courts will always do what the planners want. In many ways planning law is anomalous; magistrates and judges don't like telling people they can't live on their land, and are more conscious of human rights than our planners were. You might have a battle, yes, but you might win. More people do win these days, and permaculture is becoming accepted as a rigorous discipline for a low impact lifestyle that planning inspectors can say yes to.

3. If you want to do it the slow way, you need to recognise that unless there is some kind of policy to support low impact development in your planning area, you probably won't get permission without a great struggle, publicity campaign, appeals etc. A planning inspector might, if you are lucky, be receptive to your permaculture design and ideas and give you temporary permission, but my conclusion is that it is no more difficult to change or influence local planning policies through the established channels, which involves contributing to the consultation process that all planning authorities are obliged by law to carry out. Unless your authority is particularly enlightened, you will find its information releases in quite small unattractive legalistic blocks of small print at the back of your local newspaper. Once you are on their lists, though, they consult you about everything, even though the section across the corridor is doing everything it can to

have your house demolished, and the Masons want you stamped out as a revolutionary menace to stable land holding everywhere. Sometimes I want to laugh and cry at the same time, but it's just the way it is. To get them to listen to your ideas seriously, form an association of low impact people in your area to influence planning policy. Maybe as part of a Transition Town strategy. We had what was called the Pembrokeshire Environmental Forum. This was a group of about thirty good people who, among other events, organised a whole day on Low Impact Development. Simon Fairlie gave a presentation, with lots of local councillors and officers invited and attending and breaking into small discussion groups, etc. A few months later the draft Low Impact Development policy was published. One clause excluded the National Park. We, and lots of other people on the consultation list, including the Welsh Assembly, questioned this exclusion clause. They had to do a u-turn, but put in lots more hoops and hurdles. The final version, *Supplementary Planning Guidance - Low Impact Development Making a Positive Contribution*, can be downloaded as a from www.pembrokeshire.gov.uk. Love it or hate it, it is a policy supporting low impact developments. It might be a breakthrough. The masons might squash it. Time will tell.

4. Learn how to run a website, and put lots of pictures on it. Buy a good camera and learn how to use it. Enjoy the media. They love eco-houses, because a nice little grape-covered den makes so much better pictures than wrecked cars and courthouse exteriors, which is what they get to film most of the time. Make friends with them. What big lenses they have! How efficient they are, and busy! How the presenter's hair shines! Offer them raspberry wine or a cup of tea and sympathise with their workload. Make space in your life to show local people round and to listen to them.

5. Last but not least: Design your house to be so pretty that only an Orc would really want to pull it down. Love it and talk to it and make it your Space of Love.

That's it. 'Bye.
Tony Wrench
November 2007

Appendix

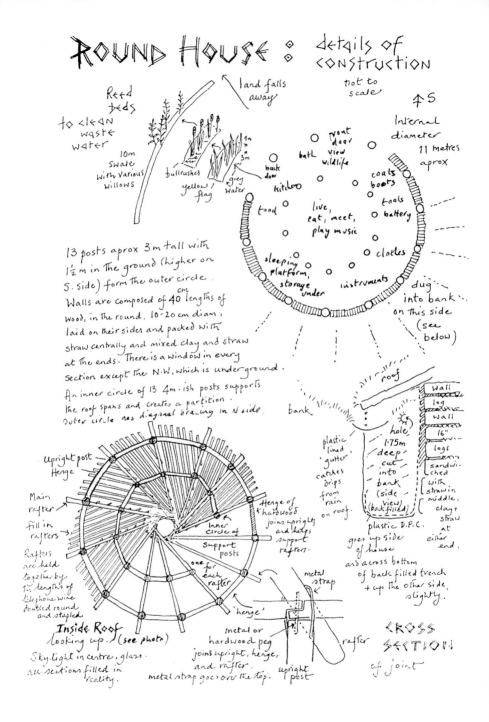

CROSS SECTION ROUND
WALL/WINDOW HOUSE

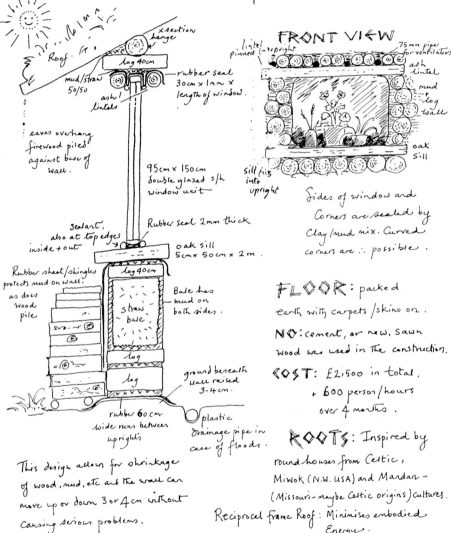

FRONT VIEW

Roof.

log 40cm

mud/straw 50/50

ash! lintels

x section hinge

rubber seal 30cm x 1mm x length of window.

eaves overhang firewood piled against base of wall.

95cm x 150cm double glazed s/h window unit

sealant, also at top edges inside + out

Rubber seal 2mm thick

oak sill 5cm x 50cm x 2m.

Rubber sheet/shingles protects mud on wall, as does wood pile

log 40cm

straw bale

Bale has mud on both sides.

log

log

ground beneath wall raised 3-4cm.

rubber 60cm wide runs between uprights

plastic drainage pipe in case of floods.

lintel pinned to upright

75 mm pipe for ventilation

ash lintel

mud + log wall

oak sill

sill fits into upright

Sides of window and corners are sealed by clay/mud mix. Curved corners are ∴ possible.

FLOOR: packed earth with carpets/skins on.

NO: cement, or new, sawn wood was used in the construction.

COST: £2,500 in total. + 600 person/hours over 4 months.

ROOTS: Inspired by round houses from Celtic, Miwok (N.W. USA) and Mardan~ (Missouri - maybe Celtic origins) cultures.

Reciprocal frame Roof: Minimises embodied Energy.

No waste: Rubber under walls
Mud on walls
Wood in fire.

This design allows for shrinkage of wood, mud, etc and the wall can move up or down 3 or 4cm without causing serious problems.

by Olwyn Pritchard

A2

CAT (Centre for Alternative Technology)

Machynlleth, Powys SY20 9AZ

Tel: 01654 702 400 Email: info@cat.org.uk Web: www.cat.org.uk

A good source of advice on alternative technology. They produce books, factsheets and tipsheets on a wide range of subjects. The site itself is well worth a visit.

Ecological Design Association

The British School, Slad Road, Stroud, Gloucestershire GL5 1QW

Tel: 01453 765 575 Email: charlierite@clara.net

Further Reading

Alternative Housebuilding

Mike McClintock; Sterling Publications; 1989

ISBN: 0806969954

Good explanations of building with logs, poleframes, cordwood, and stone. Packed with diagrams and technical information.

Complete Book of Cordwood Masonry Housebuilding

Rob Roy; Sterling Publications; 1992

ISBN: 0806985909

An inspirational as well as a practical manual combining the techniques described in the author's previous books *Cordwood Masonry Homes* and *Earthwood*.

Complete Book of Underground Houses*

Rob Roy; Sterling Publications; 1994

ISBN: 0806907282

This book focuses on earth sheltered buildings rather than buried ones. Advice ranges from selecting a site to building a complete home. Case studies, diagrams and photos plus a US resource list.

Cordwood Building - the state of the art*

Rob Roy; New Society; 2003

ISBN: 0865714754

An thorough, articulate and knowledgeable explanation of this unique and earth-friendly method of construction. From the first examples to the latest techniques, a one-of-a-kind reference.

* *Titles available from The Green Shopping Catalogue, see page A6.*

Guidelines for Landscape and Visual Impact Assessment; Second Edition
The Landscape Institute and the Institute of Environmental
Management and Assessment.
Spon Press; 2Rev Ed edition (18 April 2002)
ISBN: 041523185X
Provides advice on assessing the landscape and the visual impacts of development projects. Describes baseline conditions and significance of impacts.

Lifting the Lid - an ecological approach to toilet systems*
Peter Harper & Louise Halestrap; CAT Publications; 1999
ISBN: 1898049793
Complete with triumphs and disasters of real case studies, this book should answer all your questions on compost toilets.

Off the Grid - an ecological approach to toilet systems*
P Allen & R Todd; CAT Publications; 1995
ISBN: 1898049092
How to design install and look after small scale renewable energy systems, this is a practical guide for the DIYer, self-builder.

Our Ecological Footprint - reducing human impact on the earth
Mathis Wackernagel & William Rees; New Society; 1996
ISBN: 086571312X
Cuts through the talk about sustainability and introduces a revolutionary new way to determine humanity's impact on the earth.

The Owner Built Home
Ken Kern; Owner Builder Publications;1992
ISBN: 0686312201
Out of print but absolute gold dust if you can get a copy. There is a photocopy version of *The Owner Built Home Revisted* (ISBN:9992883154).

Permaculture Magazine - solutions for sustainable living*
Permanent Publications
ISSN: 09765663
Published quarterly for enquiring minds and original thinkers everywhere. Each issue gives you practical, thought provoking articles written by leading experts as well as fantastic permaculture tips from readers.

* *Titles available from The Green Shopping Catalogue, see page A6.*

Shelter*
Bob Easton & LLoyd Khan; Shelter Publications; 2000
ISBN: 0936070110
A celebration of handmade human habitations through the ages, from caves, caravans and lodges to timber frame, cob, adobe, yurts and domes.

Short Log and Timber Building Book
James Mitchell; Hartley & Marks; 1985
ISBN: 0881790109
A useful reference book if you can find a copy, as is the 1986 edition by Rodale Press (ISBN: 088179009).

Solar Water Heating - a DIY guide*
Paul Trimby; CAT Publications; 1999
ISBN: 1898049114
How to build and install your own panels, including a guide to materials, UK suppliers and further help.

Supplementary Planning Guide
Low Impact Development Making a Positive Contribution
Joint Unitary Development Plan for Pembrokeshire.
The essential guide for anyone considering low impact sustainable living in Pembrokeshire. Details all the conditions which need to be met.
Can be downloaded in pdf form by using the link:
www.pembrokeshire.gov.uk/ObjView.asp?Object_ID=2700.

Tapping the Sun - a guide to solar water heating*
Brian Horne & Peter Geddes; CAT Publications; 1999
ISBN: 1898049173
An introductory booklet that takes the mystique out of solar water heating in the UK. Includes a comprehensive resource guide.

Tomorrow's World
Duncan McClaren, Simon Bullock & Nursrat Yousuf
Earthscan Publications; 1998
ISBN: 1853835110
Seeks to explain what we can do to live comfortably within what we actually have through efficiency and sufficiency.

* *Titles available from The Green Shopping Catalogue, see page A6.*

Other Books in the Simple Living Series

Tipi Living*, P Whitefield. **Sanctuary***, E Edwards, et al. **Eat More Raw***, S Charter. **Getting Started in Permaculture***, R & J Mars. **Do It Yourself 12 Volt Solar***, M Daniek.

Useful Websites

Brithdir Mawr sustainable community – **www.brithdirmawr.co.uk**

Creating innovative earth-sheltered space – **www.earth-sheltered.com**

Centre for Alternative Technology – **www.cat.org.uk**

Chapter 7, sustainable living planning issues – **www.tlio.org.uk**

Communications for a Sustainable Future – **csf.colorado.edu**

Development Centre for Appropriate Technology – **www.azstarnet.com/~dcat**

Diggers and Dreamers – **www.diggersanddreamers.org.uk**

Green building supplies – **www.recycle.mcmail.com/grebuil.htm**

Philosophy for ecological architecture – **www.nacul.com/nacul.htm**

Permaculture Magazine – **www. permaculture.co.uk**

Permaculture Association (Britain) – **www. permaculture.org.uk**

ThatRoundhouse – **www.thatroundhouse.info**

This book was kindly supported by

Jay Abrahams, Bringsty, Worcestershire. Andrew Brockbank, Norwich, Norfolk. Diane Chapman, Bracknell, Berkshire. Nick Davis, Llangrannog, Ceredigon. Alan Dearing, Cas Blaidd, Pembrokeshire. Ian Ford, Carol Dines, Stoke on Trent. Dundee. J Johnson, Exeter, Devon. Suzie Jones, Robertsbridge, East Sussex. Adrian Leaman, London. Stefan Proszynski, Narberth, Pembrokshire. Richard Northridge and Bijon Sinha, Tregynon, Powys. Lysana Robinson, Swindon, Wiltshire. Karl Stirland, West Bridford, Nottingham. David Woodland, Hatlestrand, Norway.

*Titles marked * on pages A3 - A6 are available from The Green Shopping Catalogue along with around 400 other titles on related subjects. For your free copy please contact:*

Permanent Publications, The Sustainability Centre, East Meon, Hampshire GU32 1HR, UK.

Tel: 01730 823 311 Fax: 01730 823 322 Email: info@green-shopping.co.uk or order online at: www.green-shopping.co.uk